C0-AVZ-918

WHEAT FLOUR MESSIAH

Eric Jansson

OF BISHOP HILL

•

PAUL ELMEN

Published for the
Swedish Pioneer Historical Society
By
Southern Illinois University Press
Carbondale and Edwardsville
Feffer & Simons, Inc.
London and Amsterdam

Library of Congress Cataloging in Publication Data

Elmen, Paul.
Wheat flour messiah.

Bibliography: p.
Includes index.
1. Jansson, Erik, 1808–1850. 2. Jansonists.
3. Bishop Hill, Ill.—History. 4. Jansonists—
Biography. I. Title.
BX7990.J3E45 284′.1′0924 [B] 76-28380
ISBN 0-8093-0787-1

Printed in the United States of America
Designed by David Ford

TO THE MEMORY OF MY FATHER

Once upon a time there was a saga which wished to be told and led out into the world. . . . It was still only a strange medley of tales, a formless cloud of adventures, which drifted back and forth like a swarm of stray bees on a summer's day, and did not know where they might find someone who could gather them into a hive.

—Selma Lagerlöf, "A Saga about a Saga"

Contents

List of Illustrations

Preface

Bishop Hill was drowsing under a warm summer sun. My imagination carried me easily from the scene which lay before me to the same village street in 1850, when colonists who knew only a few words in English were there. It was not difficult to see, swinging down a rutted lane, the figure of Eric Jansson, the wheat flour salesman who ignited several of Sweden's northern provinces with his passionate beliefs. Twelve hundred sober countrymen left everything they had known and loved in their native fields to follow this man on the antic adventure he proposed. Some of the very buildings constructed by the colonists survive as though their builders had just left—the Steeple building, with its single hour hand, the Bjorklund Hotel, the harness shop, and several others. Some of the objects worn shiny by their hands are preserved in cases on the first floor of the Colony Church—a snuffbox, hand tools, a knife for letting blood. A descendant of one of these immigrants who believed that one could be entirely free of sin was riding a power mower over the common. In the coffee shop, people with faces like those painted by the colony painter Olof Krans (some of whose portraits are found herein) were eating hamburgers and french fries. Time seemed transfixed at Bishop Hill, like a fly trapped in a cake of amber.

It occurred to me that such anachronism should be explained. The Bishop Hill story had been examined from many points of view—and economists, sociologists, anthropologists, and linguists had thrown light on this bit of the American past. But no one had studied the colonists' primary beliefs with any care. The result was a curious distortion, because the Janssonists were religious men and women in a sense which our own age finds difficult to comprehend. Someone, I decided, should tell the story of their faith and offer some account of the Lutheran state church and its hostility, of conviction that moved audaciously into deed, of the birth and death of a dream. But could anyone give a conceptual account of a people who were sometimes stridently anti-intellectual? Even with rich resources and professional techniques, could one reach across a long century and understand people like the Janssonists? Could one with the best will in the world give a plausible explanation of their subtlety and naïveté, their delicacy and brutality, their feelings of arrogance and inadequacy? It would be hard for someone in these times to recover the terrifying simplicity of the Janssonists. But the problem was tantalizing. It seemed clear to me that the key to the Janssonists was their religious faith, and the secret of their faith was hidden with Eric Jansson, their leader.

There are clues, though the past will keep many of its secrets. The libraries of the Chicago area yielded the required printed materials. Valuable manuscript notes are in the Philip J. Stoneberg collections at Augustana College, Rock Island, Illinois, and in a similar collection at Knox College, Galesburg, Illinois. There are a great many documents at the Bishop Hill Heritage Association archives, and at the County Courthouse in Cambridge, Illinois. There was good hunting at the Illinois State Historical Society archives in Springfield, Illinois, and in the National Archives in Washington, D.C. In Sweden my main sources were the University Library at Uppsala, the Royal Library in Stockholm, and the Stockholm City Archives. Many details came from the Uppsala *Landsarkiv*, the Härnosand *Landsarkiv*, and the Emigration Institute at Växjö.

There are too many people to thank properly, but it would be churlish not to mention the help given me by Dr. Wesley M. Westerberg of the American Swedish Institute in Minneapolis, Dr. Nils W. Olsson of the Swedish Council of America, Dr. Franklin D. Scott of Claremont, California, and Miss Edla Warner of the Bishop Hill Heritage Association. The help given me by my wife, Gretalyn, and my daughter, Lisa, must remain a family secret. In Sweden my chief indebtedness is to Gunvor Jansson, Skensta, Fjärhundra, who is a descendant of Carl Lindewall, Eric Jansson's younger brother; Erik Nilsson, the chief of police in Hudiksvall; Eric Trolin, sometime editor of *Hudiksvall Tidning;* Emil Erickson of Berga, Söderala; and Olof Ohlsson, farmer and cultural historian of Oppsjö, Delsbo Parish.

With such friends to call upon and with a suitcase of notes, I went to Dalsland, Sweden and to Taormina, Sicily, to test the coherence of my tale. Somehow, I thought, it must be possible to understand even so palpably a fire-clothed man. "What song the Syrens sang," wrote Sir Thomas Browne, "or what name Achilles assumed when he hid himself among women, though puzzling questions, are not beyond conjecture."

Paul Elmen

Evanston, Illinois
August 1975

WHEAT FLOUR MESSIAH

The Visitation

> Once, during harvest, I was seized by an attack of rheumatic fever, fell to the ground, and lay limp as a rag.
>
> —Eric Jansson, "Autobiography"

Landsberga is a small village just off the main road which runs down from Uppsala to Enköping in Sweden's Uppland. The area is the nation's oldest ecclesiastical setting, and some of the names on today's map recall the ancient high-church past: Biskopskulla, Biskops Arnö, Skokloster, Sigtuna. A few miles northeast lies the ancient university, where learned men with rounded shoulders could be seen processing into the cathedral with its twin spires. The air is heavy with traditional memories, order and obedience, intelligence and sanctity living happily together.

But the village of Landsberga looks westward, too, this time to the province of Västmanland and a different landscape of fertile fields, red cottages, and rural charm. The farmers of Västmanland had their own dignity, though it was quite unlike that of the scholars and churchmen of Uppsala. It was here on December 19, 1808, that Eric Jansson, the founding father of the Janssonist movement, was born. We have his unpublished autobiography,* which opens with a sentence showing that humility was not des-

* The location of the original manuscript of Eric Jansson's "Lefnadsbeskrifning till 1844" is unknown to me, and this important document may no longer exist. There is a partial copy in the archives of the Bishop Hill Heritage Association, but there is a more complete transcription in the Herlenius collection of manuscripts at *Carolina Rediviva*, the Uppsala University Library, Uppsala, Sweden. I have quoted from the latter copy, and refer to it throughout as "Autobiography." Except where otherwise indicated, the translation of all Swedish manuscripts and printed materials is my own. I have quietly changed the spelling of place-names to conform with current usage, as defined in *Svensk ortförteckning* (the Register of Swedish Place-Names.) Eric Jansson dropped the extra *s* from his name when he came to America, but for the sake of consistency I have kept it.

tined to be his controlling virtue: "I know that after reading my first words you want to ask if I was born into the world without the agency of man, but came instead by the action of the Holy Spirit on a woman. To which I answer that it was not so. I was conceived and delivered in Biskopskulla, on December 19, 1808." [1]

One would hope that he meant to be funny, but nothing in the record or in the memory of man recalls the slightest evidence that Eric Jansson had a sense of humor. It is likely that he was writing for his disciples, who really did entertain the idea that he was more than human. After reassuring them of his essential humanity, he goes on to recall some extraordinary events of his childhood. He was harshly critical of his parents. Because of their consuming interest in worldly things, he said, they did not give their children proper attention. When he was one or two years old, he was left in the care of his brother, Jan, who was then only seven himself. While playing with a knife or an ax, Jan cut off the first two fingers of Eric's left hand. When Eric was eight, he was allowed to drive a horse and wagon. The horse bolted, overturning the wagon and flinging Eric to the ground. He lay unconscious in the road, blood streaming from his head, and he suffered frequent headaches from that day until the year 1830.

But neither the circumstances of his birth nor his early life were in any sense exceptional. He followed his parents to the church services at Biskopskulla, entering through the narthex with its wall paintings of great grinning devils, stalking through the world and seeking someone to devour. They became in time as vivid in his mind's eye as they were for Martin Luther. When the other young people reveled in each other's company without thought of man's cosmic destiny, he was never able to escape from these prowling devils. Anna Maria Stråle remembered that on July 13, 1834, she and Eric were attending a local dance and they spent the evening discussing theology.[2] When the family moved from Biskopskulla, Eric was confirmed at the nearby church in Torstuna.

In 1830, when he was twenty-two years old, an event happened which must be considered decisive for his later career. He had been troubled for many years with rheumatism, which he said was caused by his father, who "piled on too much work," and for which no remedy could be found. During the harvest time of that year, he was seized with an especially acute attack and fell to the ground with his face next to an old plow. This is his account.

> I began to think about what Jesus Christ had done when he was on earth, and when it struck me that he wanted to do the same great works today, and was able to do them, I prayed that he would make me well. At that very moment I was completely freed from my pain. I saw that up to that time I had neither sought nor found my soul's health according to God's will, I had been deceived in my trust in God and in the atoning grace of Christ, and I had been deceived in all my crying to God according to Romans 7 that I would do his will, when I had no power to do it.
>
> I realized that I was really under the law and under condemnation, as long as I carried on adultery with the Bible like a woman who is intimate with another man while her husband was still alive. I was terrified when I realized that hell was my destiny as long as I knew only the law and its demands, and did not walk in newness of life without sin through faith in the atonement and resurrection of Christ—or in other words, live by every word that comes from God's lips. I had known that the wages of sin is death, and I had tried to follow the dictates of my conscience and stay away from obvious sins, but I fell into sin nevertheless, because I lacked the power of the true faith.
>
> It dawned on me that I had been deceived in the faith which I had received from the so-called evangelical Lutheran teaching, which had been passed on to me by my parents and by various clergy and teachers. Now I knew that there was no reason why I should have lived from my childhood with a troubled conscience and an anxious soul. Why could I not before this have lived at peace? And what should I do now? I had no one to turn to, since I now knew that all the preachers and teachers were blind leaders who could not enlighten men concerning spiritual power which they had neither seen nor felt.[3]

The vision on the Landsberga barn floor was an epiphany comparable to the vision of Paul on the Damascus road, or Wesley at the Aldersgate mission. Jansson spent the rest of his life amplifying the essential features of the vision, adding very little to its substance. From now on he would know with absolute certainty that the source of all physical and spiritual healing was Jesus Christ. Wounded souls could find out about this healing in only one place: in the Bible. When one had the true faith, anything at all was possible, even though in the eyes of men insuperable difficulties appeared. Why was this vision not widely shared? The reason so few knew about it was that the established church officials had betrayed their trust and were blind leaders of the blind. There was only one hope for him and for the world: "I must let the power of Jesus Christ be in my heart through belief in his word." He picked himself up from the barn floor, dusted off his trousers, and began a new ministry which was to change the lives of many thousands of people.

At that time he took an active part in the *läsare* conventicles. The *läsare* were lay readers, encouraged by the more evangelical clergy, who liked to meet in small groups in various homes to study the Bible, discuss certain devotional books, and pray together. They translated their theology into a rigid life-style, looking askance at such nonbiblical activities as drinking, card playing, swearing, and dancing. Before a Separatist impulse took over later in the century, their aim was to cultivate individual piety and by this means bring life into the Lutheran formularies. It was a form of devotion deeply appealing to Eric Jansson, who was hungry for spiritual seriousness.

One of the books widely discussed in the *läsare* meetings was written by Johan Arndt (1555–1621), the German Lutheran mystic and follower of Melanchthon who had long been venerated by the Pietists. L. P. Esbjörn's Swedish translation of *Vier Bücher vom Wahren Christentum* had been published in Gävle in 1843, and there were not many *läsare* cottages which did not have *Sanna Christendom* [True Christianity] as a well-thumbed possession. Arndt's central theme was that soul-satisfying faith was not to be found in the utterance of learned men, though many foolish Christians looked for it there. True Christianity was to be found only in Christ. "Everyone now seeks out learned men," he complained, "from whom he can learn art, speech, and wisdom; and no one wants to learn gentleness and true humility, as well as holy living, from our only teacher, Jesus Christ." [4]

Such teaching struck a responsive chord in Eric Jansson, who had no learning of any kind and could find very little nurture from university-trained clergy. But in Arndt there was another emphasis less to his liking, because it stood athwart the plans he was beginning to formulate of having a lay ministry. Arndt, like his teacher Melanchthon, wanted to rescue the Lutheran reformation from abstract theology and turn it instead toward everyday living. He therefore warned laymen not to believe that the only way to be religious was to do what monks and nuns and professional religious did. Rather they were to be content with their calling, infusing what had been called secular work with an evangelical content. Above all they were to recognize the sanctity of the word and to be proud of being devout laymen.

Such talk was rubbish in the ears of Eric Jansson. He wrote: "I read much in Johan Arndt's *Sanna Christendom* because he had the reputation of being both truthful and also Christian. But this devilish book deceived me completely, and foolishly. I fell into the sin of no longer preaching. Arndt told me that I should remain quietly in my calling and not aspire to become a teacher. I, who was a farmer, should not be presumptious enough to try to preach. I accepted this doctrine, especially since I was fed up with the daily criticism which was being heaped on me for preaching." [5]

But now he had a problem: what about the commission to spread the Gospel which had been given to him when he lay on the barn floor? He managed to quiet his conscience with a curious subterfuge: "I began to think that there must be teachers in the world whom I did not know, and so it was not absolutely necessary for me to preach the Gospel of Christ to the people."

What Eric Jansson took to be Lutheranism's hostility toward lay preaching continued to be one of his chief criticisms of the established church. The basic position which Arndt elaborated had been laid down by Luther in his sermon on 1 Cor. 7:20, published in his *Kirchenpostile.* According to this view, each man had been given a *Stand*, or station in life: he is a husband, or wife, or daughter, or son. These orders were laid down by God. The call of the Gospel comes to each man not in general terms, but in his distinctive *Stand.* A man was certainly not to give up his *Stand* when the Gospel call came. He was to remain a fisherman, or a farmer, or a housewife—*im Beruf.* Protestants were not to make the same error which had befuddled the Catholic centuries, namely, that salvation had anything to do with one's status in life. Let the monk have his cell. The Christian was called to what seemed a more colorless but what was really a more demanding task: to be a man among men, hardly distinguishable from other men, except that he was *im Beruf.* "Remain in your station in life," said Luther, "be it high or low, and continue your vocation. Beware of over-reaching." [6]

In the early 1830s, Eric Jansson was eager to be a faithful *läsare* according to the old model, obedient to the discipline of the Lutheran church, not for a moment overreaching. The combined pressure of the local criticism and the two leading teachers of Lutheranism was too much for him, and for four years he did not preach at all. He decided that his station in life was to be an obedient layman, perhaps a *hemmason* (a son who lives at home), quiet, studious, and taciturn. But his fate was to be quite otherwise.

Though his parents were by no means rich, they had, like all similar farm families of the time, *pigor* and *drängar,* who helped with the duties of the household and farm, even though their pay was nothing more than their food, shelter and bed, and minimal clothing. One such *piga* was Maria Kristina Larsdotter, "Maja Stina," who had worked in Eric's home for six years.[7] In 1835 she was twenty years old, while Eric was twenty-seven. They were married in that year. In his autobiography, Eric mentions almost in an aside that his wife was pregnant when they were married: "We were united according to the will of God," he wrote, "before we lay in bed together. The children of the world knew that she was pregnant on our wedding day, and so thought we had done what is forbidden by men. We knew that all such talk was against Colossians II." [8]

Eric's father and mother, children of the world, were indifferent as to what Colossians, Chapter 2, said on the matter, but were distressed to know that their son had married beneath his station and as a consequence not of calm decision but from a moment of indiscretion. Eric reports bitterly that on his wedding day, Maja Stina was not allowed to join her husband at his father's table. There is reason to believe that his later frenetic rebellion against established authority was in some degree caused by the fact that he left home without his father's blessing and was forced to work out a solitary destiny outside of what Luther had called "the orders of creation."

He began preaching again the year of his marriage, though he denied that the two events were related to each other. Maja Stina was one of his early converts. The young couple faced some harrowing early years together. Johannes, the son who was born four months after their wedding, died at the age of six months from *bröst feber* ("chest fever"). Their second child, Lovisa, was born on May 27, 1837, and died three months later from whooping cough.[9] Both events were recorded in the church records, but not in his autobiography, either because they were too sad to speak about, or because he had already begun a process of detachment from the victories and defeats of ordinary life and so felt no great sorrow at all. Did the infants not go home to God? In any case, he was not so detached from worldly concerns as to be a poor farmer, but worked mightily and wisely, and in 1838 was able to buy the small property, Lötorp, Sånkarby, for one thousand *riksdalers*.

He mixed his talent for hard work with a certain bourgeois shrewdness and sold whatever he could in local markets. One such trip occurred in October, 1840, when he and his younger brother Carl took a load of animals and farm produce to sell at the market in Uppsala. He was shocked at the sinfulness of the town, the drunkenness, the profanity, the whoredom, the general atmosphere of moral abandonment. When they had sold their wares, Carl asked Eric why he had such a sad expression. "Our business has gone very well, but I can see in your face that something is wrong." Eric said he would explain on the way home. As the now empty wagon bounced along the gravel road, through Läby and Söderby and Ramsta, Eric preached a very long sermon, the thesis of which was that the world was in a general state of decay, and that the best one could do now was to have a solitary invulnerability, like Joseph in Pharoah's house. A man could be honest even in a den of thieves.

After they reached home he called on the young Österunda *läsarepräst* (pastor who was also a Reader), Johan Jacob Risberg, and presented his problem.[10] He felt that he had the power to preach and had a message to give, but both Luther and Arndt had said he should remain in his calling. What should he do? Risberg answered at once that he must continue to preach. "I am glad that you came to me," he said. "I wish all the people in the parish were as gifted as you. But one thing worries me."

"What is that?"

"That you will be overcome by spiritual pride. An unusual amount of grace has been granted you."

Eric was to hear that shrewd insight again, but neither now nor later did it make any impression on him. He said that he did not live in order to please himself, but to please Him who died for us. He said that any good he could accomplish would not be his own doing, but God's.

The two became very close friends. Risberg said that the love they felt for each other was like the love between David and Jonathan, and Eric wrote in his autobiography that it was stronger than the love of a man for a woman. For ten years they studied together, prayed together, and attended house meetings together. Their theology was much alike. For example, they both argued against the teaching, bruited about at the time, that souls would grow in grace after the body died. Eric was pleased at a sermon he heard Risberg preach attacking the doctrine of Purgatory.[11]

At one of the house meetings in Klockaregården, Österunda, Eric gave a long and passionate defense of his position that God must first purify us from sin before the Holy Spirit can work in us, citing Rev. 15:8,9, as his proof text. Both Risberg and Carl Christian Estenberg, the young man who was to be his successor in the Österunda church, happened to be present. They were astonished at this layman's power. Both agreed that though they had been to the university and had spent much money at their studies, Eric was a better soul-winner than either of them. Here was a new recruit who would waken the sleeping church! At a later meeting in the Eklund home at Sånkarby, Eric prayed long and fervently. Estenberg said that he had never felt such ecstasy, and that he hardly knew if he was in the flesh or not.[12] The heady possibility struck them that a revival could begin just here and spread throughout the church.

It is difficult to give a coherent account of the rift which developed between Risberg and Eric, because the version of it given by Eric is cryptic. Eric hints that Risberg was jealous of the attention he paid to his wife. On one of his visits to the Jasson farmstead, he told Eric that he should never have relations with his wife except when they wanted children. Eric approved of this dictum, and convinced Maja Stina of it as well. But after a while Eric began to wonder if the command had really come from God or was only one of the contingent decisions which the church was always making and then pretending that they were absolute. Here is his puzzling account: "After a while we found that all the laws we had followed were really made up by men. Because of the advice of Risberg, my wife began to doubt that she was under grace, and came into such darkness that she began to condemn herself. My own judgment of her was harsh. I strengthened her sense of being lost when I told her the more she cried and prayed to almighty God, the hotter would be the flames prepared for her in hell. I did not understand that the *läsarepräster* were the cause of her troubles." [13]

The so-called *läsarepräster* responded in the same way to Eric Jansson: at first they welcomed him as an ally, hoping that he would bring new life into ancestral forms; but they soon discovered that he had no such intention and was, in fact, a Separatist and revolutionary. The new associate minister who arrived in Nysätra and Österunda in March, 1841, was not for a moment deceived by the Janssonists. Nils Abraham Arenander set out to expose the sinister quality of the new movement. He developed the conventional preparation for Communion into a kind of Inquisition. Some fifty years after it occurred, Wilhelmina Westerberg, who was then living in Bishop Hill, put down as much as she could remember of such an examination by Pastor Arenander.

> ARENANDER: Do you intend to receive Holy Communion?
>
> WILHELMINA: Yes.
>
> ARENANDER: But you have no sin to confess, I suppose.
>
> WILHELMINA: "If we say we have no sin, we deceive ourselves, and the truth is not in us. If we confess our sins, he is faithful and just to forgive us our sins, and to cleanse us from all unrighteousness." I don't know what sins could remain after that. "Though your sins be as scarlet, they shall be as white as snow; though they be red like crimson, they shall be as wool."
>
> ARENANDER: You have the Holy Spirit then as well, my child?
>
> WILHELMINA: "Now if any man have not the Spirit of Christ, he is none of us."
>
> ARENANDER: You are a strange people. I have never heard

anyone talk much about the Holy Spirit before Eric Jansson came along.

WILHELMINA: That is surprising. The Bible says that Jesus breathed on his disciples and said, "Receive ye the Holy Ghost: Whose sins ye remit, they are remitted unto them; and whose sins ye retain, they are retained."

ARENANDER: (loudly) Get out of here! You shall not be allowed to receive Holy Communion! [14]

Fifty years is a long time, and even though Wilhelmina Westerberg had a good memory, a word or two could have dropped out, or been added, or lost its position, and the meaning subtly changed. But the conversation sounds authentic. The Janssonists knew their Bible, at least in the sense of having apt passages to use in a debate. And Wilhelmina had the familiar assurance of the Janssonists. She remembered that she did not feel the slightest awe before her pastor, and she remembered besting him easily in the discussion. It sounds right that Arenander should finally lose his temper when this impudent girl claimed for herself the power of loosening and binding, the sacred power of absolution which is one of the gifts of ordination. These insolent laymen were challenging the church's claim to be the sole judge as to who would belong to its fellowship—the very right which Arenander was at the moment exercising.

Wilhelmina remembered also that another girl from the Janssonist circle, Ulrika Andersson, daughter of Anders Andersson, had run afoul of Arenander and the church council. Ulrika, who may have been rehearsed for the interview, quoted Col. 2:13–14 when the pastor asked her about her notion that her sins had already been forgiven and needed no further absolution: "And you, being dead in your sins and the uncircumcision of the flesh, hath he quickened together with him, having forgiven you all trespasses; blotting out the handwriting of ordinances that was against us, which was contrary to us, and took it out of the way, nailing it to the cross." Arenander had forbidden her to receive Communion the next day, but Ulrika came to the altar rail anyway, waiting to receive. When Arenander passed her by, administering neither bread nor wine, Ulrika had to return to her pew. But she was not subdued. That afternoon she reported triumphantly to her friends: "I have had a heavenly dinner today.

Jesus says, 'Whoever will permit it, I will go in and sup, I with him and he with me.' "[15]

The colloquies with Wilhelmina and Ulrika illustrate the Janssonist method of dealing with the clergy. The method was to quote Bible passages, so that the authorities were made to appear in violation of Scripture itself. The assumption that the truth could be held as a simple, uncluttered possession gave them a self-confidence which knew no measure, and which refused to defer to the agents of the government. The latter assumption was what infuriated Arenander, rather than the theological differences. He wrote to the officer of the Crown in Västerås: "That we are all doomed to hell by them, especially the undersigned, does not matter very much. But surely it is too much when one is attacked insolently in one's own house. I thank God that I have been given a strong arm which can shove them from my door."[16] And when he signed Ulrika's permission to leave for America in 1846, he wrote in the church records some incoherent words which showed he had not yet simmered down: "Burned Arndt. Paradise. The Garden of Eden."[17]

The mounting hostility in Västmanland promised no great future in that province for the Janssonists. Risberg, who had been born in Gävle and who admired the *läsare* piety of Hälsingland, agreed that his friend would be wise to take a load of wheat flour for sale and head north to meet some friends who would not greet him with Arenander's heavy sneers. Eric was invited by a friend, Anders Arquist, the associate minister in Enånger, Hälsingland, to come up for a visit, and Johan Risberg wrote a warm letter to insure his welcome.[18] Eric had reason to believe that he would find friends in the north, and at the very least sell a few sacks of wheat flour.

The Sale of Wheat Flour

Around January 1, 1843, I travelled to Hälsingland to search out my brothers in the faith.

—Eric Jansson, "Autobiography"

Two figures in a sled pulled by an old white horse moved slowly along the rutted road which lay between Gävle and Hudiksvall—the same which later generations would number E 4, and call "Europavägen" (the Europe road). The sled was hung with sacks of wheat flour which were to be offered for sale to the housewives of Hälsingland, who had bins of rye flour but who longed for bread as white as their famous linen, and which could be made only with flour made from wheat.[1] The accepted farm lore was that the fields of Hälsingland were too far north to make wheat-growing a sensible undertaking. The man who held the reins was of average build and height and was wearing a blue coat. A wisp of light brown hair protruded from his cap. His face was thin, and his cheekbones protruded. There was nothing distinctive about his whole appearance, except that his eyes were singularly intense, as though he burned with some ravaging conviction that had nothing to do with wheat flour. His lips were thin and straight, and when he spoke he revealed uncommonly large front teeth, which gave him the appearance of a rabbit.[2]

Though no one along the coast road knew it, Eric Jansson was making an ominous entry into Hälsingland. Since the eighteenth century, this part of Sweden had been a breeding ground for an assortment of religious enthusiasts. A hundred years before Jansson arrived, lay zealots from Lillhärdal had come up along the coast with apocalyptic messages, and they had been joyously received in Söderhamn and Hudiksvall. There were Pietists and Herrnhutists reading their Bible under the lamps in red cottages. Lately a strong party had sprung up under the leadership of a Finnish pastor named Hedberg, who had rejected the rationalism of the church handbook of 1811, and had pleaded for a return to an older Lutheran piety. The American Presbyterian Robert Baird had been enthusiastically received there, and so had the English Methodist George Scott.

The *läsare* held their meetings in homes despite the Conventicle Edict, which since 1726 had prescribed fines, imprisonment, and exile for those who met secretly without the leadership of the church. But until the 1840s the *läsare* were quiet, devout laity, who did not propose at all to separate from the Lutheran church but rather to fill it with warmth and power. Hälsingland had been untroubled by the Conventicle Law since 1779 when Svante Myrins led the battle for tolerance. The state officials looked the other way, and many of the clergy were sympathetic with them and even joined their services. There were many clandestine meetings to discuss their favorite authors: Luther, Arndt, Nohrborg, and Tollstadius. As early as 1737, Knut Lenaeus had written to a brother clergyman that his parishoners in Delsbo "study Tollstadius' writings and believe in him as though he came from heaven." [3] In the middle of the century the servant girl Karin Jonsdotter had preached at excited meetings around Söderhamn. There were devoted followers of Jacob Boehm and also of Emanuel Swedenborg.

The soil was right for the kind of seed Eric Jansson intended to plant. When he came to Ina, not far from Söderala, he pulled up his horse and asked some men on the wayside if there were any *läsare* who lived in the neighborhood. They said there was one at Ina, not far away. Eric proposed to his companion, Petter Jansson, that they spend the Sabbath there.

Thus occurred the historic meeting between Eric Jansson and Jonas Olsson, the farmer who was widely known as a leader of the Hälsingland *läsariet.* [4] Without official title or status, he was a natural leader of men whose word carried great weight throughout Hälsingland. Not knowing yet that Eric Jansson was dead in earnest, and was never known to tell or recognize a joke, Jonas asked him if any good could come from Västmanland. [5]

On Sunday morning, Jonas explained that there was no *läsarepräst* in Söderala, but he intended to go to church anyway. The local clergyman at that time was Anders Scherdin, who had been in charge of this flock since 1824. He had a deep evangelical interest, being a member of the society "Pro Fide et Christianismo,"

Jonas Olsson (1802–98), farmer and *läsare* pastor from Ina, Söderala Parish, whose acceptance of Eric Jansson paved the way for Jansson's successes in Hälsingland, and who later became one of the colony leaders. Painting by Olof Krans, courtesy of the Bishop Hill Heritage Association

and he was active both in the "Evangelisk Sällskap" and in the "Svenska Missions Sällskap." [6] But Jonas was right that he was no *läsarepräst*. In February, 1846, when the Cathedral Chapter at Uppsala was making a belated effort to understand what had caused the rise of Janssonism, they asked the provincial governor, Baron L. M. Lagerheim, to gather information. He wrote to all the clergy in Hälsingland and asked them questions about Janssonism. Scherdin wrote back to him that the *läsare* were responsible for the rise of Janssonism. "*Läsariet*," he wrote, "can certainly be seen to have laid the foundation and prepared the way for Eric Jansson's heresy. the *läsare* paid more attention to their own assemblies than they did to the church worship, and had more time to complain about their pastor than to seek his instruction." [7]

Jonas Olsson told Eric Jansson that he did not wish to antagonize Scherdin, so he planned to attend the service, and invited Eric to go along. That afternoon the *läsare* gathered in Jonas Olsson's home for the more serious worship, and Eric was an interested spectator. He reported that one C. G. Blombergsson, a noncommissioned officer (*fanjunkare*) in the Hälsinge regiment, opened the meeting by reading a postilla.[8] Then Jonas Olsson preached. Eric watched the *läsare* and heard them sigh and groan. The young wheat-flour salesman from Västmanland was preparing his attack against this kind of piety, and wondering how his own theology could be most effectively presented. These people seemed to him to have a genuine desire for spirituality: they were distressed by their sins, and they cried aloud for God's grace. But they had no saving power. They seemed to him always to be seeking, without ever finding, always pleading for forgiveness, but never feeling forgiven. The time had come, as he put it, for the sword of the Gospel.

Still he hesitated. Jonas finished preaching and then, perhaps only to be polite, introduced the visitor from Västmanland, who had, as he said, "made great strides along the way." Eric resented the patronizing friendliness and later on he wrote: "He [Jonas] thought he was far ahead of me in knowledge of the Scripture and ability to preach. He treated me like a child who needed some fatherly support in spiritual matters. I thought of 2 Cor. V:12, and that I should perhaps take it easy with these people to win them over. I approached Jonas Olsson like a servant who had something important to tell his master." But, he said not a word after being introduced, and Jonas closed the meeting with a prayer.

He decided to confront Jonas directly.[9] There had been no prayers for the household, he said, during the time he had been there. He realized that older people have trouble listening to advice from the young, but there was something he simply had to say: Jonas was not being a father in God to his people or to his family. "You still do not know the Lord's way. Your children do not obey God's word, nor do your household. If you cannot manage to be the head of your own family, how can you expect to be a leader for other people?"

The question struck Jonas like a blow, and for a moment the fate of Hälsingland hung in the balance. If Jonas had reacted with anger, driving this brash young salesman from the house, the Janssonist movement would have had great trouble getting started in Hälsingland, and may not have been begun at all. But the moment passed, and Jonas capitulated. Eric had exploited the secret knowledge of all spiritually sensitive people: that all men are unworthy servants, and that in God's eyes no man living is justified. And Jonas jumped from this admission to another that seemed its consequent: that Eric Jansson could save him and bring this perfection for which he longed. He admitted to Eric that though he had been a *läsare* preacher for sixteen years, he had accomplished nothing at all and could be said to have lived in sin. He longed for the kind of dedication obviously felt by the young visitor, who, when Jonas's sister Carin came over to buy some wheat flour on Sunday afternoon, told her that a Christian did no worldly business on the Sabbath. For the rest of Jonas's long life he was a follower of Eric Jansson.

Moving northward along the coast, Eric stopped at Norrala, but here his reception was different. He met another lay reader, P. Norin, but he did not meet the local pastor, Lars Olof Norell, who was probably busy in his workshop. No ordinary carpenter, he was honored in the community for having built a wagon and two pianos.[10] The Methodist missionary from England, George Scott, considered him one of his best friends in Hälsingland. The lay reader, Norin, would have nothing to do with Janssonism, and his reward was to have a taste of Eric's invective: "You stand and warm yourself with the Atonement, just as Peter stood by the fire in the priest's palace, But you deny the Lord in your lives." [11] Norin ushered him to the door of his house and warned him not to come back. Jansson summarized his view of the city:

"Norrala is like a burnt-over hillside on which nothing can grow." "The judgment is hard," said Norin, "but coming from Eric Jansson's lips, it doesn't matter in the least." [12]

Moving north on the coast road, he came to Enånger, where he met a mixed bag of Swedenborgians, followers of Strauss, and some old-fashioned Lutherans, but he also met two *läsarepräster*, Johan Olof Nordendahl and his associate Anders Henrik Arquist, both of whom welcomed him.[13] At Njutånger he met Olof Stålberg, who greeted him as an "angel of light," and admitted ruefully that Jansson had received more light from the Scriptures than had ever been granted to him.[14] Stålberg recommended that he continue along the coast until he came to Hudiksvall, where he could meet Schoolmaster Claesson.

Clas Eric Claesson was an ordained Lutheran clergyman who taught many generations of Hudiksvall students in the local folk school. He together with Sefström and Norell were very close friends, sharing deep affection for "Bror" Scott, an evangelical piety, and a sympathy for free church movements within the church.[15] But there was not much other Janssonist strength in this bustling lumber and fishing port. Its pastor, Carl Ludvig Boström, reported to Baron Lagerheim in 1846 that Hudiksvall had been spared Jansson: "Among Hudiksvall's 2,000 inhabitants there is only one old fisherwoman who is known to praise Eric Jansson. The woman has been known as a *läsare*, but one fly does not make a summer. . . . I think that such sleek swindlers [as Jansson] can never sink their roots in a congregation where the pastor tries with will and ability to enlighten, lead, and help his flock.[16]

On his way home, Jansson stopped at Hälsingtuna, where he met a redoubtable opponent, Anders Ersson of Hasta. Jansson told about the reception at Norrala, whereupon Anders said, "Don't go there, because if you do, Norell will pull the mask from your face." [17] Anders quizzed him about the doctrine of sinlessness:

"Do you believe, Eric Jansson, that a person while still in this life can be free from sin?"

"He should try to be."

"Yes, I know he should, but the question I asked was whether a man can ever reach such a stage of freedom from sin."

Jansson said that a man might sin once, but he should never sin a second time, because if he does it becomes "uppsåtlig" (intentional sin).

"How long," continued Anders, "have you been living without sin?"

"About three years."

"Three years is a long time—more than a thousand days. You must be an expert in holy living. If one is tempted daily, you have in three years confronted and overcome more than a thousand sins. Do you have any sins left?"

"I have not yet fully overcome my doubts."

"No, that answer won't do. Doubt is the mother of all the sins. How does doubt show itself in you?"

"Either you or I will go to hell." [18]

On the way south he met Jonas Olsson, who decided to travel with him as far as Gävle. Together they sought out a *läsarepräst*, who arranged a meeting for them. But when the people assembled, it was Jonas who was invited to speak, to Eric's chagrin. On the whole, however, the Hälsingland visit was a success.

Maja Stina was pleased to see him back at Lötorp at last. Pastor Risberg called to get a report of the trip and was especially pleased to hear that the *läsarepräster* had welcomed him. Eric spoke at local *läsare* meetings, and his tone was now much more confident. Some of his listeners were moved by what he said, but they thought he went too far. They could go along with his theology, but why did he demand absolute perfection? "I told him that God made even more strong demands than I. You know what history teaches us of John Huss: he gave in a little, and was immediately burned at the stake."

Led by an obscure but powerful impulse which he did not understand himself, Eric Jansson chafed to return to Hälsingland, though his wife strongly opposed the idea. "Neither she nor I," he wrote," believed that I would leave her defenseless for such a long time with the care of Eric and Tilda." [19] Relatives were, of course, critical, and blamed Maja Stina for the difficulties she was in. With curious detachment, Eric reported what some of his own family had said to her: "You deserve to suffer, and no one will help you with the children or with the chores, because you should have prevented your husband from going on this trip." But, there were forces here beyond Maja Stina's control. Eric decided to make a second trip to Hälsingland, this time, because it was summer, on horseback.

The first stop now was Gävle, where Jansson met again the *läsare* leader P. Estenberg, Risberg's friend. He had intended to preach, but he had no sooner begun than the *läsare* snatched the Bible from his hand and began his own sermon. It lasted a long time, Eric thought, but he waited patiently until the end, and then preached his own sermon, this one, he said, having true power. The Gävle episode was a hint of the growing resistance. At Hille and at Trögås he met nothing but hostility. At Hamrånge he met the Reverend Johan Eric Fillman, who was at first friendly but who later developed into one of his worst enemies.[20]

Wryly he noticed that the reverse happened at Söderala. Katrina, Jonas Olsson's wife, first met him with open hostility, and later became one of his most staunch supporters.[21] Jonas Olsson, who was now completely captivated, saddled his horse and followed his friend northward. As they trudged along the snowy roads, they discussed the strength and the weakness of the *läsare.* Jansson had discovered both their longing and their frustration. P. N. Lundquist, himself a *läsarepräst* from Maråker, wrote the first book about Janssonism and made the same point that Eric made in his talk with Jonas Olsson: "Holiness was not a pearl which was sought [by the *läsare*]. There was an eagerness to hear the Word preached, but not to achieve faithfulness, or to do faithful deeds, or to hear the Word with love for the truth." [22] Jonas had found a friend who was readying some new weapons with which he could overthrow the Lutheran church and the *Ecclesiola*, the *läsare* church within the church.

The Reverend Olof Stålberg greeted him at Enånger. He had a reputation as a Pietist, and had in fact been a leader in the "Evangeliska sällskap" (Evangelical Society). Much criticized by his flock, he had been the subject of a visitation by Archbishop Johan Olof Wallin, but the prelate had in the end supported him.[23] Stålberg said that he felt as though the ground shook beneath him when Jansson preached. He cried like a child, and said he envied the power that Jansson had been given. But there were no other followers at Enånger or Njutånger, and even Stålberg began to have second thoughts. "His wife," wrote Eric, "led him to destruction." Jansson left what he called "this Babylon" and came to Hudiksvall, where he stayed at the home of a friend of the colporteurs, Per Hermansson.[24]

At this point he was joined by a young girl from Delsbo who

was to have a harmful effect on his Swedish ministry. Her name was Karin Ersdotter, from Nyåker, Delsbo Parish, and everyone called her "Bos-Karin." Jansson was careful to put into his written account that he and she often prayed together and had many deep discussions of spiritual matters. She had sat beside him when he was composing hymns and had heard him sing some of them for the first time. Somewhat feverishly, he wrote that if she told the truth she would have to bear witness to his piety: "I say that if she does not admit that God was with me, the very stones would cry out!" When his voice was hoarse after long hours of preaching, and when his legs felt as though they could no longer hold him up, Bos-Karin was at his side like a faithful friend.

At Forsa he preached twice. While there he was asked if he would go north to Bjuråker, and since Bos-Karin knew the area well, she went along as guide. At that village he met his "friend and brother in the faith," Per August Åkerlund. Together they discussed the best way to approach the influential minister from nearby Norrbo, Anders Gustaf Sefström, who had already promised that he would come to hear Jansson preach.[25] Sefström's approval would help, since he was well known for his work in poor-relief and the temperance movement, was well liked by all, and was sympathetic to *läsare.* Like Lars Norell and Olof Stålberg, he was a member of the George Scott coterie. In fact, his English Methodist friend had helped him to publish a book, *Några blad til historien om läsare, med afseende fästadt på de inom Helsingland vistande* [Introduction to the history of the readers, with especial attention to those in Hälsingland (Falun, 1841)]. It was important for Jansson to win Sefström over, and he and Åkerlund discussed how they had best approach "the great dragon in his lair."

They decided that the best approach would be a frontal attack on the *läsare,* the kind Jansson had mounted before Jonas Olsson at Ina and before Claesson at Hudiksvall. After the service, the people had started to leave for their homes when Sefström rose and made a confession very like that made by Jonas Olsson: "I confess before God and this servant of the Lord as well as before all of you, that I have taught you to walk along a way on which the Lord is not to be found. I have wandered in darkness and unbelief, and none of us has found the power in Christ which we longed for. I ask you all to receive this light which God has sent to us. He that despises Eric Jansson does not despise a man, but God and his Holy Spirit!"

The next day Sefström and Åkerlund accompanied Jansson to Delsbo. Standing on the road between the church and what might be the loveliest bell tower in Hälsingland, Sefström confessed that the devil often led him into doubts. Jansson comforted him with the promise that he would remember him in his prayers. But he also remembered that Lars Landgren lived in the rectory nearby and hastily left what he once called "this house of devils." He went to Nyåker, Bos-Karin's home.

At Forsa near the end of March he seemed near physical exhaustion. Suffering from a severe toothache and finding that his legs seemed to buckle under him, he lay upstairs hoping to get some rest. But the people had begun streaming into the farmyard and were eager to hear the famed evangelist, though some had also come to refute his heresies, and some were simply curious. The tension revived his drooping spirits, and he spoke for over four hours on the terrors of hell prepared for all those who believed in anything other than the Gospel he proclaimed. Many believed him, while others went away shaking their heads.

On March 17, a soldier named Snål from Trögsta asked to come to his house, and one of Jansson's friends, Corporal Käck, showed him the way there.[26] The meeting turned into a debate. The *läsare* quoted Luther, who said that we are never free of our sins until we die. "You ought not to put Luther's word ahead of God's word," said Jansson. "You can't squirm away from 1 John III, or Gal. II:17–18." He told them that though they spent a lot of time calling upon God, they had still not been given the power to forgive sins. "But only God can forgive sins," they replied. Jansson said that their devilish teachers were all blind. "If the Lutheran clergy," he said, "put on even more albs and chasubles than they already have on their shoulders, and if the Catholics put a lot more crosses on their churches, and decorated their altars, and drank a lot more wine than they do now, they would still be unable to forgive sins!" He said that the Lutheran clergy complained about the Catholics giving absolution for sins in exchange for money. But were the Lutherans any better off not forgiving sins at all? The more he prayed and wept and argued, the angrier Snål got, and finally he threw open the door and told Jansson to be on his way.

Bos-Karin and others went with him to Hudiksvall, where they stayed at the house of their friend, Schoolmaster Claesson. Claesson had received a letter from Sefström which heartened them. In it Sefström had reaffirmed his tribute to Jansson, and had written, "You have been a tool in God's hands." The letter moved Jansson deeply. "Reader," he wrote in his "Autobiography," "I am moved to tears when I think of all the *läsarepräster* who have broken bread with me, and knelt at my side, and asked me to pray for them."

But as happened with distressing frequency in Jansson's life, Sefström's effusive compliments turned very quickly into abuse. What changed the pastor's mind does not appear in written record, though it may have been when Sefström noticed the Separatist tendencies which lay only partly concealed in Jansson's theology up to that point. Or it may have been personal. Sefström's wife disliked this wheat flour salesman, and she spoke her mind: "How can you stand a man like that?" she demanded. "He stayed with us for three days, and during that time he never once took the saddle from his horse's back." [27]

Jansson seemed more at ease with Sefström's criticism than with his compliments, since he was declaring war on the Swedish church, including the *läsarepräster*. In a gossipy mood, he wrote: "Bos-Karin told me that he [Sefström] had seduced a girl who was about sixteen years old, and since then had found other ways of appeasing his desires, which could not be satisfied by his own wife. His lust was soon revealed to the world, since the girl gave birth to a child. Because the miserly Sefström would not give the girl the money required in such cases by Swedish law, she went to him and asked him to help support the child. Now at last he was frightened—but for what? That his preaching tabs would be taken from him. Sefström agreed to her proposal and then bribed the county councilor so that the whole thing could be hushed up."

All evidence is not available on library shelves, and it is possible that the story is true. But it must be said that all the evidence which has survived points the other way. Sefström had his critics at the time, but they had nothing to say about his sexual morality. The newspaper *Aftonbladet* complained only that he was a Methodist.[28] Surely it is significant that *Svenska Missions-Sällskap* made him inspector of schools in Lappmark, 1840, and that he was a leader in the *Svenska Bibel Sällskap*. None of these positions is perhaps incompatible with adultery, but his reputation among his contemporaries seems to have been as high as that expressed by

George Scott, who called him in a letter to Archbishop Wingård "the estimable Sefström." [29]

The meeting in Hudiksvall on March 17 was at Fisherman Skoglund's house. Jansson described the local pastor, Carl Boström, as "leaven from hell." But a later disciple, Johan Lundquist, the papermaker from Bergvik, Söderala Parish, was also there.[30] Someone whom Jansson called "a Pharisee" asked him to preach at his house outside the city, but the man turned out to be interested only in hearing something new, and Jansson and his party returned to Hudiksvall.

Jon Olsson from Stenbo came riding up. He had three sons who had heard the debate at Snål's house on March 17 and had been convicted of their sins. Since then, they had sought for some saving light, but none came. Would Jansson go back with him and counsel his sons? Eric harnessed up his horse and drove away with him, riding all night until they arrived at Stenbo in the morning. That day, March 20, he preached his first sermon in Stenbo. The sons were gloriously saved, as were also Jon Olsson and his wife. But the parents believed only for a short time, Jansson claiming that their greed was too strong and overcame their spiritual needs.

A revival broke out in the neighborhood, but there was opposition as well. Karin Jonsdotter from Utnäs traveled about the community and held meetings at which Janssonism was denounced, but she was not able to unite the opposition. At the end of March, Jansson left Forsa for Söderala, together with a party of about twenty, including Jonas Olsson, the printer C. G. Blombergsson; his sister Lovisa Hård, from Nektjärn, Mo Parish; Olof Jansson and Jacob Jacobsson, both from Valla; the tailor Nils Hedin, and Fru Hebbe.[31]

He returned home at the end of April to find that disaster had struck. Thieves had broken into their home and had taken away most of their worldly possessions, leaving his wife and children in abject poverty. The house meetings had turned into eerie gatherings, some aged women speaking in hushed voices of angel dances, and of dead saints which had returned and spoken to them in their wine red houses. Worst news of all, Risberg was now openly unfriendly, concerned no doubt because he was soon to be transferred to another parish by a worried hierarchy.

The disaffection of Risberg was accomplished by ecclesiastical pressure. Anders Johansson, the learned pastor in Giresta who had won a prize from the Swedish Academy for his translation of Juvenal, sent a letter on the situation to the Cathedral Chapter at Uppsala.[32] He said that the Österunda *läsariet* had been growing in size and frequency of meetings, and that Pastor Risberg visited them and led them. The Torstuna *läsare* were now joining the Österunda group. When Pastor Nordien of Nysätra and he had met with Risberg in the Nysätra rectory, Risberg had argued that the *läsare* movement was not dangerous. On the contrary, he thanked God because his people were now awakened from their sleep. He reported that when he had attended their meetings, he read nothing more dangerous than the Bible, or Arndt's *Sanna Christendom*, or Nohrborg's or Petterson's homilies. This kind of meeting was surely better than the usual dancing and drinking which went on while the church looked the other way.[33]

Johansson quizzed Risberg about Eric Jansson. He asked if Jansson had made several trips to Hälsingland. Risberg admitted that this was true, but that he could not prevent his going. Johansson warned him of the danger of giving free rein to "a man with stirred up fantasy, but without appropriate education to develop his understanding." [34] Risberg agreed and promised that he would do all in his power to stop Janssonism. Ten years of friendship between Jansson and Risberg were at an end.

The World, the Flesh, and the Devil

> We certify unanimously and at one time that the traveling wheat flour salesman Er. Jansson from Sånkarby in Österunda, presently in Västmanland, confessed in our presence this winter that he had by means of shameless talk tried last summer to seduce the servant girl Karin Ersdotter from Nyåker, which shameless proposal this girl rejected with abhorrence.
>
> —Signed by six witnesses at Delsbo, May 6, 1844

Despite the opposition from the local clergy, Jansson set out on his third trip to Hälsingland, this time passing through Gästrikland to northern Hälsingland. He had a two-wheeled cart, the kind used for hauling things about a farm, and both sides were hung with sacks of wheat flour.[1] He would drive his horse and cart to the farmhouse door, tell about his flour, and then strike up a conversation about the soul's welfare and the possibility of holding a meeting that evening. In places like Söderala and Älfta the news of his coming spread rapidly through the farmhouses, and a large crowd was certain to gather for the meeting.

At Ovansjö he stopped at the home which he said belonged to "a rich church warden." This must have been Erik Grip, son of Hans Ersson, the owner of the local mine, and whose wife Katarina Jansdotter was also the daughter of a mine owner. They were all part of a large crowd of Janssonists which left Gävle to sail for America on October 14, 1845.[2] At the crossroads town of Bollnäs, where the Bible scholar F. O. Söderlund presided over a handsome church which had been dedicated in 1468, Jansson found takers for his wheat flour but no believers in his special theological insight. Swinging north to Järvsö he was rudely questioned by some of the church leaders, among others the formidable Pila-Karin from Våga, who established the local line: "You can be sure that we know very well what you are up to. You are

Pehr Jansson (1815–64), brother of Eric Jansson, from Sista, Torstuna Parish, who became one of the seven trustees of the colony. Painting by Olof Krans, courtesy of the Bishop Hill Heritage Association

not inwardly what you seem to be on the outside. Get on your way, and don't come back!" [3]

This hint of a disreputable background to his evangelism was to pursue Eric Jansson throughout his years in Sweden, and in fact has persisted to the present day. There were never wanting people who said that his claim to sinlessness was a theoretical position at best, and that he had more than the usual share of fleshly appetities. His critics around Forsa speak knowingly about the "Jesus baby" said to have been left by "Ersh-Jansa," or as Vilhelm Moberg called him, "Vetemjöl Jesus" (Wheat Flour Jesus).[4] At Söderala he was publicly admonished by Karin Olsson, sister of Jonas and Olof, because of his alleged improper relations with a married woman in the parish.[5]

Some of the charges are clearly malicious, brought by people who had something at stake in blackening his character. And they were plausible because of the passion of his speaking, his air of emotional excess, his indifference to established laws and customs. Suspicion also arose because of his manner of travel. He seemed always to be moving about Hälsingland without his wife and in the company of several women, eating and sleeping wherever they happened to be. In the fall of 1843, he traveled around Bollnäs in the company of "Fugda-Britta," an unwed Älfta girl, and "Bröd-Marta," daughter of the baker from Bollnäs. "The madam and the Bollnäs girl," said P. N. Lundquist, who had looked into the matter, "spent nights in the same room with Eric Jansson, and many saw with disdain the relationship between these three persons." [6]

This kind of thing was too much for Lars Landgren, who had come to Delsbo as pastor in 1842. Born in 1810 in a farmhouse at Östervåla, Västmanland, not far from Jansson's birthplace, he was one of the founders of "Evangeliska Fosterlandstifelsen" (Evangelical National Foundation), and not one for philandering. Tall, heavily built, with a deep commanding voice, he was widely respected in the neighborhood and called "Lång Lasse i Delsbo" (Long Lars from Delsbo). He had been forgiven for behaving in a very unprofessional manner during his first ministerial assignment at Delsbo: sobbing as he led a woman who had been condemned for murder to the gallows.[7] But there was no faltering when six members of the Delsbo church brought charges against Eric Jansson, and the traveling evangelist was brought to trial on November 18, 1845.

The charge was that Jansson had attempted to seduce a twenty-seven-year-old young lady from Delsbo, Karin Ersdotter, known as "Bos-Karin." [8] The incident occurred at Svartvallen, a *fäbodvall* ("summer pasture") deep in the forest south of Delsbo where local farmers in rotation brought their cattle during summer months to feed on the rich grass in clearings under the brooding pine trees.[9] Only one of the original cabins now remains, but in the summer of 1843 there was a village of seven or eight of them, with an equal number of barns nearby for the cattle. They were manned by *pigor* and *drängar*, young men and women whose job it was to take care of the cattle, milk them mornings and evenings, and then separate the cream from the milk. The assignment was looked forward to by the young people, who found it a change from the routine of the home farm, and who had plenty of time for socializing after the chores were done for the day.

Bos-Karin was staying at the cabin which belonged to the family from Nyåker shortly after midsummer, 1843, when Eric Jansson came through the door, together with Per Forss, who was cutting timber nearby.[10] Karin knew Eric well, having met him in Hudiksvall the previous winter, and she traveled together with him on various preaching trips throughout Hälsingland. Her report at the Delsbo church trial was that Eric had come in smiling and had said to her that she would be his next wife. Jansson spent the night with Eric Forss, and Karin slept in a cabin with Anna Olofsdotter, the servant from Sunnansjö.

The next day Bos-Karin was alone in a cabin when Eric came in. She said that he used enticing words again and made it plain that he felt strongly attracted to her. His exact words were that he wanted to "göra dig en lill stinta" ("give you a little baby"). Shocked, she refused. He said he could do anything without sin. Then he added, "If I wanted, I could bewitch you" (the words are hard to translate: "toka bort dig"). She replied sarcastically that God's power is greater than Satan's power. As he was leaving he told her that she would soon be overcome, but she was not to tell a soul about this conversation. If she did, it would be considered an offense against God Himself.

On the third day they had prayer in one of the cottages belonging to Kjerstin Andersdotter, Karin's sister-in-law, where Jansson had been invited to sleep.[11] Afterward they talked far into the night, and Karin lay down on the bed and fell asleep. In the morning she woke up and found Eric Jansson under the blanket at her side; "Oh, oh, Kjerstin," she said to her sister-in-law, "what have you done to me?"

Six young people, one of whom, Brita Pedersdotter was Eric Jansson's disciple, on May 6, 1844, presented a formal indictment to Pastor Landgren charging Jansson with having tried to seduce Karin Ersdotter.[12] The charge seemed well substantiated, but Eric Jansson was not shaken. He would not admit that he had made any indecent proposals to the girl. What he had really said was this: "If I should say to you that I wanted to have relations with you, do you think you could be seduced?" In other words, he was testing her to see the strength of her faith. Concerning the "toka bort" remark, his explanation was that he had really said, "Do you know what your lot would be if I had 'tokat bort dig'? Yes, you would go to hell." What about his remark that she should be his wife? He had simply meant to pay tribute to her: she had followed him faithfully on many trips, she had cared for his welfare, she had been in fact like a solicitous wife to him.

Karin would have none of this version. She told the court some more things that happened: a few days later, when she was lying in bed, Erik Jansson came and sat down beside her. He said to her, "Never before have I wanted to get rid of my wife. But now I do. If I had met you earlier, I would never have married my wife." Jansson said he could not remember any such remark. His answer to the claim that he slept with her was simply that he had indeed talked until a late hour, and when he finally lay down on the bed which had been assigned to him, he found Karin on it. But he had kept the blanket between them and he never touched her.

Did he not think it was wrong to do this? At that time he did not think it sin to lie beside a woman without making any demands on her, he replied. But now he was aware that it was probably against God's will. One should not only be innocent but should avoid even the appearance of evil. He agreed that it would have been much better if he had slept on the floor.

Eric Jansson made another curious attempt to justify his conduct in the Bos-Karin affair. When he was leaving Sweden in 1846, he sent a copy of his farewell to Sweden and a covering letter to an unknown friend giving him power of attorney and some ammunition to use in Jansson's defense when he might be tried in absentia before the Cathedral Chapter of Court. Jansson wrote:

"You have a reputation of being cruel and outrageous to individuals and therefore you claim that a righteous man can sin if he wants to. That's how it was with that Delsbo girl, whom I, according to 1 Cor. V:5 ['to deliver such a one unto Satan for the destruction of the flesh'], have handed over into Satan's hands because she didn't keep her teeth in front of her tongue. Supported by Ruth III:14, I asked her to sleep until morning at my feet, but she told everybody about it, exactly opposite to the explicit commands of the verse ['And she lay at his feet until the morning: and she rose up before one could discern another. For he said, Let it be known that the woman came to threshing floor.']." [13] In Jansson's autobiography, he presents still a third version of the episode to prove that it was all Bos-Karin's fault.

> I had an idea that Bos-Karin felt physical desire for me. I knew that in fourteen years I had not felt any sexual temptation, and so I had reason to think that whatever was happening between us was her fault. To find out whether this was true. I deliberately made myself guilty. I told her that I had a desire to possess her, though not to live with her. Within me I felt that she had spiritual power, but this struggled against the knowledge that she was burning with desire for me—a desire that could not be restrained, much less fulfilled. I told myself that she should test whether she was free of guilt as I. I told her to face that she was full of desire for me. She denied it. I decided that I had better leave.
>
> I went to another hut where they had told me to stay. Before I began preaching that evening, Bos-Karin came and admitted that she felt desire for me. Afterwards she threw herself on the bed where I was supposed to sleep. I felt very tired from my travels, and had preached until after midnight. There was no other bed for me. Anyway, I lay down on the same bed where Bos-Karin was sleeping, but separated from her in all ways. [14]
>
> If this confession should be read on the Judgment Day, I know that all the other letters and papers about this affair are lying. I have admitted that I felt natural desires, but I did not do what David did. I did what Scripture says we should do: I killed my fleshly desires. Bos-Karin, like Potiphar's wife, was frustrated in her lust for me and then she said it was my fault. [15]

If the Janssonists had trouble swallowing this explanation, they did not let the world see their doubts. In Söderala at the end of May, 1844, after Jansson said that he had been without fleshly desires for three years, Olof Olsson, who was not nearly as steadfast a disciple as his brother Jonas, spoke up: "Wait a minute, how about the Delsbo girl?"

"Oh, yes," replied Jansson, "that girl in Delsbo. Well, that was punishment from God, because I believed that Magister Sefström was a true Christian, and that was a terrible sin." [16] We do not know if Olof Olsson was able to sort out the wild logic of this answer, but the main body of followers did not even try. They were satisfied to know that "even though Eric Jansson did this and that, his heart was nevertheless righteous before God, no matter what his body does." [17] And they had cause for complaint that no one of the Janssonists, who had lost his citizenship when he fell out with the church, was allowed to testify at the Delsbo trial, thus insuring only hostile witnesses. "The six witnesses summoned to Delsbo," they wrote in an 1846 tract, "could not testify that he had done anything other than he said, namely to test Karin Ersdotter. And the one who of the six witnesses told the whole truth was kept out." [18]

The reader may be forgiven for being unwilling to decide whether the Joseph and Pharaoh's wife story, or the David and Uriah's wife story, is the more apt biblical parallel to the Svartvallen episode. If they had thought of it in the years to come, his followers might have quoted some lines which T. S. Eliot remembered from Christopher Marlowe: "Thou hast committed—fornication—but that was in another country, and besides, the wench is dead." She was not in fact dead, though grievously invalided. On a trip back to Sweden later on in the century, Olof Fraenell, the wagonmaker, was in the neighborhood of Forsa and decided to call on Bos-Karin. He had a sad story to bring back to Bishop Hill. Bos-Karin was now an old lady who had never married, and who had been bedridden for twenty-eight years. To his surprise, she asked him no questions at all about Eric Jansson or about Bishop Hill, and left it to him to decide if she found the subject too painful to talk about, or if she had forgotten all about it. [19]

But it took a long time before Eric's enemies forgot about the story in Hälsingland. After the trial, his critics seem to have gotten the upper hand, and the Svartvallen story drifted through the countryside. Jansson's theory was that J. O. Fillman, his arch enemy from Hamrånge, had written letters spreading lies about him throughout the province. In any case, when he reached Söderala, the local *läsare* leaders Jonas and Olof Olsson told him that he had better not preach for a while. They urged him instead to go with them to hear his new enemy, Pastor Anders Sefström, preach a temperance sermon.

Neither Eric Jansson nor his followers had any great enthusiasm for temperance sermons. It was obvious enough that there was a problem of heavy drinking in Sweden at midcentury, and the ravages seem to have been particularly great among the young boys of the peasantry. It was known that some of the clergy brewed liquor from excess grain and sold brandy to their parishoners. But there was also a strong temperance program afoot. The *läsare* movement was unanimous in its opposition to intoxicating beverages, and some of the Lutheran clergy—men like Per Niklas Lundquist in Maråker, Anders Norrell in Norrala, and Anders Sefström in Norrbo—made it their central theme. But the Janssonists were out of step with the movement.

Jansson had been told by Risberg that he must be a *Nykterist* ("temperance leader"), and at a meeting in Maråker Jansson had made some mild remarks about the dangers of drink and the joys of abstinence. But he did not see the dangers as vividly as did the Olsson brothers, whose father had been a drunkard. "What did you think of Sefström's sermon?" they asked him after the meeting.

"If any man gives up drink," Jansson replied, "it is because of greed." He meant that drunkenness or sobriety in themselves were not matters that interested him. Self-centered greed was worth talking about. A man might drink out of greed, or give up drink out of greed, and be no closer to holiness because of either. The sanctification he spoke of transcended grubby ethical considerations and led to a radiant oneness in which one could be free to drink or not to drink, whatever anyone else or any other society thought.

The truth is that he used this freedom to take a nip or two himself. When Pastor Lundquist asked him about the rumors that he was not a teetotaller, he made no effort to conceal the truth. "Eric Jansson needed medicine," wrote Lundquist. "During a conversation with this author, he said that he sometimes needed a drink, since he often had a pain in his stomach which went away after a drink. I told him there must be some other way to relieve a stomach pain. He answered that he lived a long way from any store that sold medical supplies, and drink was the quickest cure available to him." [20] There is no evidence that Jansson drank excessively, and if there were any such evidence his many enemies would have exploited it.

On the whole, Jansson's third trip to Hälsingland was a disaster. At Delsbo he was told of an old lady who was chronically ill and bedridden. He went to the house with a party of followers to demonstrate his healing powers, prayed lustily and announced that the lady could now get up if she wanted to. "Holiness has now come to this house!" he cried. "He is mistaken," the old lady said, and turned toward the wall.[21]

There was no healing miracle; there was talk about Jansson being soft on alcohol, and the Bos-Karin story was on everyone's lips. It was on this trip, however, that he met and enlisted the support of two people who were later to play a prominent part at Bishop Hill. One was the tailor, Nils Hedin, a talkative and energetic little man who was to be a brisk missionary for Janssonism.[22] The other was "Fru Hebbe," who had been born Margareta Wansberg, a Söderala girl. She had married Carl Gustaf Hebbe, a saddler, and had come back with him to Ina from Stockholm in 1838. To her husband's dismay, Fru Hebbe had become an ardent Janssonist, so much so that when the *Agder* sailed in 1846, she took her four children and left with the Janssonists, while her husband stood weeping on the dock. Shortly afterward he returned to Stockholm, where he died in 1850.[23] There were some victories and some defeats during this third trip to Hälsingland.

It was now clear that the clergy would not give Eric Jansson a pass to travel to Hälsingland as a missionary, and so he had to wait until the harvest was over so that he could pretend to be selling wheat flour. He bided his time, doing the farm chores and writing a book. P. A. Huldberg, the publisher and bookseller from Falun who had shown sympathy for the *läsare*, seems to have given him some encouragement.[24] Jansson wrote to him from Sånkarby on October 4, 1843.

> Brother in Christ, and worthy friend:
>
> I sit down in stillness before going to bed, because tomorrow I have been invited to go all the way to Hedemora.

> Today I sold the goods which I had intended to take to Falun. . . . The matter we agreed on—that I should publish the manuscript if God so willed—seems about to come to pass. I have met Pastor Risberg and he has agreed to make a fair copy of it; but it has turned out longer than I expected, and other matters have hindered me, so I have not been able to finish it at the agreed time. But it should be ready by the end of the month.[25]

Carl C. Estenberg, the Österunda pastor, tried to head off this publication, no doubt fearing for the spread of views which he now felt to be dangerous. On November 1, 1843, he wrote to Huldberg with a warning.

> Brother Huldberg:
>
> I must ask you to read very carefully the manuscript which Eric Jansson may have sent to you for publication. The man disappoints me, now that I have a chance to get to know him better. He has got the idea that he can be an author, but I doubt very much that Christ's spirit and humility will be apparent in his writings. God knows! I now have his wife and children with me. We are crushed down with poverty and need. I would like to have a song printed, but I lack the money.[26]

On November 23, 1843, Eric's father, Johannes Mattson, died, and some obscure family quarrel cast an additional shadow over the family's loss. The most reasonable guess is that the father disapproved of Eric's separatism, and Eric accepted this difference with bad grace. In any case, there was an argument which marred the obsequies. This is Eric's account of the matter: "My father died and I was not allowed to come near his deathbed. Those nearest me, including my brothers did not believe in me. Estenberg, who once said that I was full of the Holy Spirit, now said that I was sent by the devil." [27]

That he had appraised Estenberg's attitude correctly is proved by a letter which the latter sent to Huldberg on November 27, 1843: "Your good opinion of Eric Jansson must not be based on my judgment. In order not to carry as a burden on my conscience the feeling of having deceived you, I wrote what I believed in the last letter. Now after I have seen how the man behaved during these days beside his father's deathbed, I am convinced that I was taken in by him when I thought well of him. He is certainly in the eyes of God a wicked man."[28]

After selling Lötorp for one thousand *riksdalers*, Jansson set on his fourth and last trip to Hälsingland. At Bollnäs and Nora he ran into fierce opposition, and on the road toward Järsvö he was suddenly struck with deep doubts about himself. Was he, as he had told his followers, a messenger sent by God? Or was he what some of his critics said, an agent of the devil? He writhed and twisted as he rode along, willing to believe either possibility. But somehow his doubts were suddenly resolved: he was God's man, sent to proclaim the true Gospel, sent no doubt to suffer. He was filled with joy, and knew forever afterward an inner peace, and never again doubted his mission. From now on he was to be gripped with the terrible simplicity of his task: the building of a true Church on earth.

At Forsa he made the decision to move up to this friendly land, where he had followers on every hand, and where he could hold services in many farmhouses. Passing through Bollnäs again, the wolves attacked as usual, but his followers rallied around him, and he left town with two ardent disciples following, Bröd-Martha and Lisette Wiberg. They made their way to Alfta, where a mass meeting was announced for Anders Andersson's house. People began streaming in from all directions around Alfta, and the house would not hold the eager listeners. A window was removed, so that people could gather in the yard and still hear what he had to say. He preached loudly for five hours without stopping, and the next day he could only whisper. The assistant pastor of Alfta, Hans Norberg, heard it all, and at the end of the day he said that there was no doubt Jansson was sent by God.

Traveling back to Bollnäs in early March, 1844, he found that some of his followers had changed their minds and were now saying critical things about him. Jacob Vadman, the Schartauist pastor from Hanebo, told people that Jansson was an agent of the devil.[29] When Jansson reached Anders Andersson's home in Näsätter, he found that he no longer had a welcome. There were still many loyal supporters around Alfta, but Pastor Norberg had now changed his tune and was warning his people of the danger they were in. The pattern was repeated many times: Jansson's fervent preaching enchanted both lay and clerical *läsare;* but many became his enemies when they discovered that his teaching was dangerous and unsound. And his personal life lent credence to the charges of decadence bruited about over Hälsingland coffee tables.

Life without Sin

> As the splendor of the second temple at Jerusalem far exceeded the splendor of the first, erected by the son of David, so also the glory of the work which is to be accomplished by Eric Jansson, standing in Christ's place, shall far exceed that work accomplished by Jesus and his Apostles.
>
> —Eric Jansson, *Catechism*

The archetypal idea which governed Eric Jansson's life and which gives his teaching in Hälsingland power and cohesion was that of perfectionism. He thought that it was possible for the believer to lead a sinless life. One of the verses which he often used to show the biblical basis of the doctrine was 1 John 1:7: "If we walk in the light, as he is in the light, we have fellowship one with another, and the blood of Jesus, his son, cleanseth us from all sin." If one accepted the radical simplicity of his thought there were only two kinds of people—his own small band of sinless believers, and the rest of the unbelieving world. His followers had seen the light, and had entered into a new and radiant life which the Catholic centuries had reserved for its saints. There is no way to understand the Bishop Hill adventure without grasping the theological presupposition: Janssonism was a dream of a pure holiness.

The Bible sets up a rather puzzling polarity on the matter, and the next verses from John, book 1, quoted above add a qualification: "If we say that we have no sin, we deceive ourselves, and the truth is not in us. If we confess our sins to God . . . he is faithful and just to forgive us our sins." How did the Janssonists deal with this obvious polarity? Simply by saying that the second half of the quotation was intended for nonbelievers. The nonbelievers should admit that they have sin, but after they have been converted and moved into the light, they need have the darkness no longer. In his *Catechism*, Jansson faced the problem head on. He has an enquirer ask, "Can you who say you are free from all

Krans's copy of a painting by Clarence N. Dobell, *From Shore to Shore.* Since no portrait of Jansson exists, it has come to symbolize his departure from Oslo. Courtesy of the Bishop Hill Heritage Association

uncleanness and free of all sins, be in need of forgiveness for trespasses or debts?" The correct answer was "Yes, the Lord's Prayer refers to the sins of other people, as Jesus himself used it." [1]

Poor, uneducated, disdained, and later hated by the establishment, Eric Jansson announced to the sober-faced people sitting in a ring around him that he was sent specially by God to preach a Gospel which had been hid from the wise and the mighty since apostolic times. In theological terms, he felt his mission was to shatter the traditional distinction between justification (forgiveness of sins) and sanctification (achievement of holiness). The two were in his mind inseparable, and happened at once. When one of his listeners was caught up by his torrent of passionate speech, and came forward afterward to enlist as a follower, Jansson simply asked, "Do you want to be holy?" When the penitent said yes, Jansson shook his hand and said, "You are holy." [2]

The idea of sinlessness is archetypal in Jansson's thought, finding its way into most of his sermons, and repeated endlessly in the songs he wrote and in the catechism he prepared for his followers. The notion of an absolute innocence provided the gravamen in his charge against the Lutheran church, the secret of his appeal to the *läsare,* who longed so desperately for this holiness, and the adequate explanation of why his followers could endure hatred, exile, separation, suffering, and death itself as though they were all incidents in the life of devotion. To his listeners sitting on square brown chairs or on rugs on the floor, the hope of sinlessness came as a dizzying possibility. The touch of the prophet's hand would banish their own sense of sin and ambiquity, and usher them into a freedom where they would be immune to temptation, innocent, triumphant.

There was, of course, the possibility of losing this grace, which the Janssonists seemed to think a decisive disaster, happening only once. When Jon Jonsson, an ardent Janssonist, was being quizzed about his beliefs at the Forsa trial on October 11, 1845, he agreed that the state of perfection was not necessarily permanent. "So long as we receive perfect grace," he said, "which is Jesus Christ himself, and pray without ceasing that he will maintain the grace and power he has given us, we cannot sin; but as soon as we do not obey his voice, we go astray." [3] When the Word is lost, the believer falls headlong from bliss and belongs at once to the Devil's party, is no longer able to understand the Gospel, and is filled with rage at those who have managed to remain in the light.

The followers of Eric Jansson were not widely read men and women, but were instead sensible and decent farmers from Hälsingland, Dalarna, Västmanland, and Uppland. They had no knowledge of religious history, and were not directly influenced by the various currents which flowed through Europe at the time. But the theology which Jansson proclaimed and which they found credible had nevertheless a long and respected history, though it would perhaps have puzzled them to have it pointed out. There are demonstrable connections between the Eric Janssonists and Eastern perfectionists. Gregory of Nyssa and St. John Chrysostom would have understood their claims instantly, though they believed that the perfection was a grace conferred at baptism. The Latin church has never fallen such an easy victim to the perfectionist mythology as has the Eastern church, but even here the medieval mystics taught that a spiritual perfection was possible in the third stage of the mystic way. The deification which emerges in the writings of Meister Eckhardt, in which the identity between mystic and God is lost in favor of a dazzling unity, resembles strikingly those passages in Janssonism which spoke of the prophet as though he were divine. And the ideal of a blessed community of those elect who had achieved perfection must be considered a Protestant parallel to the monastic ideal.

But Western Catholicism guarded itself from the dangers of a strident perfectionism in a way which Eric Jansson would have been wise to study. St. Augustine thought that perfection was not possible, except in the sense of a perfection of seeking, rather than of finding. He admired those who sought after perfection and despised those who said they had found it. St. Thomas Aquinas rescued the doctrine from *superbia* ("pride") by distinguishing between mortal and venial sins, the former of which could be eliminated, while the latter continually returned. In a stunning anticipation of impulses which emerge from below the level of the conscious mind, he thought that one could be perfectly free only of those sins which one had consciously chosen. "Man can keep from mortal sin," he said, "which is grounded in reason, yet he cannot avoid venial sins because of the disorder of the lower sensual desires, the movements of which the reason can indeed severally repress . . . but . . . cannot repress all." [4]

Jansson would have resented the suggestion that there were Catholic positions remarkably like his own, but he would have been equally shocked to hear that there is a line which leads from Renaissance humanism to his own excited discovery of human possibility. Just as the men of the Renaissance rejected the strain of pessimism which is never far beneath the surface in the medieval world view, so Eric Jansson rejected the obsession with human limitation which is one of the classifying traits of the Reformers. Jansson's faith in his own powers was inordinate. Some allowance must doubtless be made for the possibility that his assertions were deliberately extravagant for rhetorical effect, or because his opponents treated him as though he were nothing at all. But whatever the cause, there are passages in his utterance which are symptomatic of mental illness. One of his songs had not the slightest trace of native Swedish shyness: "I am perfect as God is perfect,/ The Father's life is mine as well." [5] And in his *Catechism* he expressed an unbridled messianism: "Christ's coming is revealed in its full height by Eric Jansson's obedience before God." [6]

The doctrine led him into giving an astonishing account of world history. Like many pacifists, he believed that the great disaster of Western Christendom was when Constantine came to power and changed the detached, underground religion of apostolic times into the officialdom, priests, ritual, and the chaste dullness of the Lutheran service. The world had lain in darkness between the third century and the birth of Eric Jansson in 1808. Not only that, the light which was breaking with his birth would be even brighter than that which had lit the apostolic church. He thought that his own achievements would be greater than those of Christ.[7] One would expect this type of megalomania would bring forth nothing but sympathetic sighs, but it seems in some measure at least to have been accepted seriously by his followers. Because they were utterly convinced they were able to sell their properties and move out into the unknown in order to be near him. When Maja Stina Larsdotter from Forsa Parish was asked about it she did not hesitate.

"Is Eric Jansson your God?" she was asked.

"Eric Jansson is as good as God," she replied.[8]

There is in this class of utterance something patently pathological, with its own psychological history; but there is also evident the same mood as was to be found in reacting against the pessimism of the Middle Ages and its view of the insignificance of man. In a curious way, Jansson and his followers were the unwitting inheritors of the optimism of the High Renaissance, though the humanists dispensed with the kind of grace which was always present in Jansson's thought. Both agreed that there could be a spectacular fulfillment of human possibilities here on earth.

It is tempting also to see in Janssonism a theological version of the romantic impulses which were sweeping Europe at precisely this time. The argument may seem more ingenious than sound, because Jonas and Olof Olsson did not read *The Sorrows of Werther*, nor were they aware of the cultural currents which were discussed at Uppsala and at Lund. Nevertheless, hopes had spread through Europe after the revolution in Paris in 1789–90, and in the century that followed, everything seemed changed because of the glimpse that had been given of a perfect happiness of earth. What Hazlitt called "romantic generosity" dominated politics and art. For did not the romantic poet cherish the moments of lyric intensity, and did he not invest ordinary people and objects and occurrences with an extraordinary interest, just as Eric Jansson was to do? The romantic dream of "the great good place" was to become in Jansson's followers the driving impulse toward the journey to America.

But one must agree that Roman Catholic mystical piety, Renaissance humanism, and romantic Utopianism are all ancillary forces, conditioning perhaps the atmosphere of midcentury Sweden and making more direct impulses seem more plausible to the Janssonists. For the controlling influence of Janssonist perfectionsim was his way of reading the Bible. He had learned how to give absolute priority to the Bible from Lutheranism, but the Lutheran clergy soon learned better than to argue with him by matching Bible verses, because when he was challenged, he and his followers could rattle off dozens of biblical passages like the ending of the Sermon on the Mount: "Be ye therefore perfect, even as your father which is in heaven is perfect" (Matt. 5:48). Anyone muttering something about sin, the need of repentance, or growth in grace, found himself very shortly arguing not only against Jansson but against the Bible itself.

The biblicism which is so pronounced in all Janssonist apologias was of course a hallmark of the sectarian piety which for a half century had been sweeping through the parishes of Sweden. It was a Reformation starting point, but Jansson did not follow the line of the classical Reformers, Luther and Calvin. These Reformation formulations looked for an intimate union of church and state, with religious motifs penetrating secular structures at every conceivable point. But the sectarian ideal, which dominates Janssonist thought, stressed the incompatibility of church and secular society, and counseled a continual withdrawal from all conventional and secular entanglements. "My worthy friend," he wrote to P. A. Huldberg, "pray in Jesus' name for me. And if there is anyone around who seeks the way of life without hypocrisy, ask him to pray for a poor fellow from Österunda whom God has plucked like a brand from the fires of sin." [9] The holiness sects did not mix with the world, and expressed their separation by ethical strictures such as prohibitions on smoking, swearing, dancing, cardplaying, and attendance at the theater. Jansson showed only moderate interest in this kind of legalism, but he carried to an extreme limit the notion of the sacred company, entrusted with a holy mission and of course surrounded by uncomprehending and hostile infidels..

Janssonism may best be understood as a very vigorous exploitation of certain sectarian motifs which had appeared periodically in European history. He is very like Thomas Münzer, the German Anabaptist, who in the fifteenth century led an attack against Lutheranism filing the same charges brought by Jansson in the nineteenth century. Münzer appealed to a semiliterate, impoverished following, and he held out to them the hope of establishing the kingdom of God on earth. He sacked what he considered an idolatrous cathedral, tore up and burnt as many theological books and manuscripts as he could find, and set up a primitive communism among his followers. Unlike Jansson, however, he was very interested in social problems and led the Peasant Revolt of 1525. Luther reacted violently, denounced Münzer as a tool of the devil, and urged the princes to suppress the rebellion with swords. Thomas Münzer was arrested and beheaded after a perfunctory trial, and the movement ground to a halt.[10]

Several of what Troeltsch called "the fighting sects" on the Continent resembled the Janssonists, though it is almost certain that Jansson did not know of their existence. The Family of Love and the Grindletonians could have used Jansson's *Catechism* with pleasure. So could the Brethren of the Free Spirit, who believed that if one only believed he could be perfect, he would become so, as God is perfect. All claimed to have as a single possession the key to the unconditioned meaning of existence. In England, the Ranters, like the Janssonists, denied the possibility of sinning; but they drew a conclusion from this that it did not matter at all what they did, and they went about flaunting their freedom by a radical antinomianism, drinking ale, swearing, whistling, and whoring.[11] Except for the single instance of the Bos-Karin affair, the Janssonist teaching was never antinomian, though some of their critics said that in practice the Janssonists were lawless.

The reason may be that perfectionism found its way into Janssonism not by way of Cromwell's left-wing sects, but by way of a very decorous Wesleyan Methodism. If the Bishop Hill colonists tried to explain what they believed, they often used the shortest possible way, which was to say that they were Methodists, and when the colony broke into two camps in 1860, many of the Janssonists became Methodists.[12] The distinguishing doctrine in Methodism is what John Wesley called "entire sanctification," or "full salvation," or "Christian perfection," and sometimes "the second blessing." Writing to Robert Brackenburg in 1790, Wesley spoke of it as the key to his teaching: "This doctrine is the *grand depositum* which God has lodged with the people called Methodists; and for the sake of propagating this doctrine he appears to have raised us up." [13]

The English Methodists had roughly the same charges against Anglicanism that Jansson had against the Church of Sweden. Wesley did not think that one could enter the fullness of the faith simply by being born. He had an elaborate plan of sanctification, a kind of *cursus honorem*, by which the mortal man or woman prepared for immortal life. To be sure, the Christian life begins, as the Anglicans and the Lutherans said it did, with baptism; at this stage the infant is regenerated by "infused grace," and loses his guilt for original sin (though not, unfortunately, original sin itself).[14] Because original sin survives, and is yielded to, men need to be converted, and to be filled with "prevenient grace," which opens their eyes to sin, and "convincing grace," which persuades them of their own sin. "Justifying grace" then comes with assurance that sins are forgiven. When Wesley had received this justifying grace, he was able to walk down Aldersgate Street in 1738, his heart feeling strangely warmed.

But Wesley said there was an even higher form of spirituality beyond conversion: sanctification, or Christian perfection. The positive side of this final gift is synonymous with perfect love: the sanctified love God with heart and soul, and their neighbors as much as themselves. According to Wesley the sign of the perfect man is that he is a perfect lover. The negative side of the teaching stressed the freedom in this last stage from sin, and this was the part of the doctrine which Eric Jansson settled upon. While Jansson claimed this perfection for himself, Wesley made no such claims. The founder of Methodism thought that it was possible, through a beneficient God, to be delivered from all sin, and that such perfection could be given in an instant. But he did not think that such excellent spirituality was often granted before one's deathbed, holding with St. Thomas Aquinas that this Beatific Vision was not a simple historical possibility. But Wesley did not deny that such perfection as Jansson claimed was sometimes to be seen. In *A Plain Account of Christian Perfection*, Wesley said that a person making this claim should be believed if he could give a clear account of when he received this second blessing, and if his subsequent words and acts and demeanor verified his claims. In any Christian community were to be observed people who demonstrated by the warmth of their love that they had wills entirely governed by God and never deliberately broke one of the divine commandments.

But of course even the Christians of the second blessing make mistakes, and are tempted, and even succumb to temptation, because the body still suffers from the corruption of the Fall, even though the soul does not.[15] The best of men need forgiveness. "Even perfect holiness," Wesley wrote, "is acceptable to God only through Jesus Christ." [16] The authentic Methodist perfection, which Jansson appropriated in a corrupt form, is only a relative perfection, suited to men and women still living in a fallen world. It is marked by a demonstration of love, and it claims no finality. "There is," Wesley said, "no perfection of degrees none which does not admit of continual increase." [17]

Methodist perfectionism came into Sweden by way of George Scott.[18] A young Englishman who was during his years in Sweden a revivalist and temperance preacher, he later became president of the Methodist church in England. He came to Sweden because of the circumstance that Samuel Owen had built a foundry and machine shop in Stockholm, and had brought in English workmen who were experienced in this kind of work. They needed a clergyman to hold services in English, and Owen appealed to the Wesleyan Missionary Society, which responded by sending over J. R. Stephens, and after him, George Scott. He arrived in 1830, at the age of twenty-six, full of zeal to deliver the *grand depositum* of perfectionism.

But not at first to the Swedes. Since the eighteenth century, foreigners who were not Lutherans could have their own places of worship in Sweden, but the Swedes were not allowed to visit them. The law, however, was not carefully enforced, and George Scott by the warmth of his message and geniality of temperament won supporters not only among the English workmen and consular representatives but also amongst the Swedes. He learned to preach in Swedish in one year.[19] Though he said that he had no intention of winning converts from the State Church of Sweden but wished them to remain Lutheran awakened in their faith, he nevertheless wanted his own chapel, and began in 1837 to raise money for it. During his visit to America for this purpose, he was reported by the press to have said scornful things about Swedish religious life, and Johan Ternström led a right-wing attack on him in *Aftonbladet*. When Scott returned to Sweden in 1840, an angry mob stoned Bethlehem Chapel and forced him to flee back to England.

It is not possible to establish a decisive connection between George Scott and Eric Jansson. It is clear, as Professor Mikkelsen was the first to point out, that Jansson's staunchest supporter, Jonas Olsson of Söderala, had worshipped in the Bethlehem Chapel when visiting Stockholm to sell linen.[20] It is also clear that George Scott traveled through Sweden preaching on temperance themes, and that he had large meetings in Hälsingland on at least two occasions. Though Jansson and Scott apparently did not know each other personally, the influence of Scott was widespread during the precise years when Eric Jansson was formulating his theology, and the two preached a perfectionism which is too similar to be coincidental. As had been pointed out, one whom Eric Jansson thought of as a friend, P. A. Huldberg of Falun, was a Scott disciple.

However, a search through his own writings and such records as remain of George Scott's ministry in Sweden reveals that he had no strong attachment to the doctrine of perfectionism. Certainly he did not speak of it with the same ardor as he displayed when he was pointing out the evils of brandy. Wesley's order of sanctification proved either too much for his halting Swedish, or he discounted its validity or its usefulness as a revivalist theme in the nineteenth century. In any case, his treatment of the motif was perfunctory, ending with a feeling of awe before a mystery. He wrote: "The order of sanctification, or at least as much of it as is necessary for us to understand during our time of preparation, is set forth in the New Testament; but to completely understand the operation of God's wisdom, power, and mercy, which is revealed in man's salvation, is impossible for us in the world; for in this plan are mysteries 'which things the angels desire to look into' (1 Peter 1:12)."[21]

If, as seems likely, George Scott was the vehicle by which perfectionism reached the Janssonists in Sweden, it is easy to understand why the doctrine was received in a garbled form. Wesley's account of the order of sanctification was overloaded with theological terms which only professional students of such matters would find interesting or helpful. George Scott, whatever his motives, delivered a form of perfectionism without any of the reservations which made the doctrine credible. The negative element in the *grand depositum*, stressing freedom from sin, resulted in an inevitable self-deification, and a patronizing attitude toward other forms of piety, as well as to a head-on collision with Lutheran teachings on sin. It should have been clear at the beginning that the absolute separatism required by such a theology could not be tolerated by an authority established as the Lutheran church was established in Sweden, and in that inevitable conflict it would not be George Scott or Jansson and his followers who would win out.

The outcome might well have been different had Jansson been attracted to the positive side of Methodist perfectionism, and urged his followers to outdo all others in Christian love. George Scott was himself an intelligent and kindly man, entrusted with a delicate mission in Sweden, and he had no intention of promoting separatism among his hosts. The notes he used to deliver a series of lectures are preserved in the library Carolina Rediviva at the University of Uppsala, and his exposition of Matt. 5:48 ("Be ye therefore perfect") is indicative of his general attitude toward perfectionism. This is what he planned to say: "God is Love—to be filled with pure Love—is to be perfect as our Father in Heaven is perfect—a Command—a Promise—What do ye more than others?" [22] Jansson would have agreed that perfection was a command, but would have denied that it existed as a promise.

There might have been a different story to tell about Janssonism had the Wesleyan-Scott formula become the pattern instead of Jansson's model of a perfection instantly achieved. History as arbiter in this dispute must conclude that Wesleyan perfectionism was creative, while the Janssonist version led to a tragic end. It is tempting to assert that the impulse toward social welfare in Great Britain was in some measure a product of the love perfectionism taught in the Welsh Sunday schools. On the other hand, Janssonist perfectionism had no such impressive social consequence. Sweden's concern for the underprivileged—for the aged, the young, the dispossessed, the infirm—arose out of humanitarian sympathies in the nineteenth century which were not directly related either to the established church or to the sects. Lutheranism had adopted a theological view on the mixture of good and evil in all men, and had therefore at its heart a disposition to regard sin as in some degree inevitable. But this was poor motivation for social reform. The sects on the other hand were concerned with personal salvation. Eric Jansson had little interest in social or political policy, except of course that he wanted religious freedom.

He did not define *love* as an attitude toward our fellowmen. His concern was subordinate to his desire to show his own miraculous healing or clairvoyant powers. There were, however, men and women who benefited by his mysterious power. Anders Andersson from Domta, Österunda Parish, claimed to have been healed from a serious illness simply by the laying on of the prophet's hands, just as Jansson himself was miraculously healed. And some people were helped by his power over nature. He said to Anders Andersson, a farmer from Sista, Torstuna Parish, "Get in all the hay you can as fast as you can. We're going to have such a downpour of rain that no one will remember its equal!" Andersson believed him, worked furiously to rescue his hay, and watched with astonishment a rain storm which lasted for three weeks. Skeptics in the neighborhood smiled and said it was a lucky guess, or else Eric Jansson was one clever farmer; but the true believers hid the story in their hearts as one more example of Jansson's superhuman powers. He seemed to them to be larger than life, a beacon lit by strange hands, a prophet who kept receiving special messages from a grateful God.

Righteous and Sinful at Once

> Some people make fanciful claims and say they are perfect and cannot sin any more, thus misusing the Bible and other devotional books exactly like Satan.
>
> —Johan Arndt, *True Christendom* (1606)

The Reverend Mr. Hans Norborg, interim pastor of the Alfta church in Hälsingland, was at his wit's end. One-tenth of his parishioners, the most devout churchgoers in Alfta, had become enthusiastic Janssonists and had stopped coming to his services. He wrote an anguished cry for help to the county administrator, Frederick Holmdahl. "I enclose some minutes," he wrote, "from a parish meeting: Since one often hears that agents sent out by Eric Jansson are trying to convince credulous people that our wise and gracious King thinks the official reports on Janssonism issued by the clergy and by crown authorities to be lies, let these minutes testify to our anxiety." Pastor Norborg went on to spell out his anxiety: two hundred and fifty people, he said, had left the Alfta church and had scorned the religion of the land. He had hoped in the beginning that the movement would die without official action, but he now saw that the hope was unfounded. "Boldness," he said, "grows when there is an absence of punishment." [1]

Lars Johan Hierta, the editor of Stockholm's liberal newspaper *Aftonbladet*, printed Norborg's plea and then added a sarcastic comment: "As long as the clergy itself does nothing, it seems that the secular arm has little reason to move." [2] According to this influential journal the clergy by responding to such a serious threat had lost the confidence of the Swedish people. There were hints that the reason for the timidity was that Janssonism had powerful patronage from the court, and what Norborg took to be rumor was the truth. On May 31, 1845, P. A. Huldberg, now the editor of *Hudikswalls Veckoblad* (whom Eric Jansson mistakenly thought was his friend), wrote a cryptic comment which may have referred to Norborg: "A clergyman who with power set himself against Janssonism has had to grovel." In the issue of September 27, 1845, Huldberg wrote a long article claiming that the clergy were to blame for the rise of Janssonism, "a more dangerous heresy than Strauss' faulty teaching." He thought it incredible that the leaders of the diocese did not come at once to Hälsingland and deal with this menace.[3]

It is true that the Church of Sweden did not respond quickly to the threat of Janssonism. The provocation was clear enough: a small party led by a strolling wheat-flour salesman had grown into a sizeable sect, which was meeting as a cabal and making threatening gestures toward the national church, forbidding its members to attend the common worship and describing the local clergy as devils. Machinery was available for dealing with such insolence. Why did the Lutheran discipline hesitate until it was too late to act effectively?

There are several reasons. The Janssonist movement looks different to us, having the advantage of later history, than it did to the clergy of Hälsingland in 1843 and 1844. To many of them it seemed a promising lay movement which could pour new life into forms which had begun to seem calcified. The period of rationalism had just swept by and left a debris of incredulity which made all symbolic forms seem archaic. More than one observer might have agreed with a visitor from Germany, Dr. C. Sarwey, vicar of the Lutheran church in Kirchheim, who traveled about the country in 1847 and gave his impression of the religious life: "The church of Scandinavia exhibits that firm and heavy character, in which the objectiveness of the existing relations, and particularly to its symbols, overbalances the feeling of the individual. At the present time, after a full recognition of human rights in which personal piety is not overlooked, it is necessary to take care, lest in constancy to its Lutheran confession, it overlooks or neglects the confession of the Scriptures, or even the confession of Jesus Christ in sincerity of heart." [4]

Jansssonism offered plain people who lived somewhere south of the main road an opportunity for an exciting participation in worship. More than that, it provided a way for the expression of deep feelings, for "the religion of the heart," rather than of subtle speculation. Most of all, it seemed to be at its very roots biblical, leading way back to what all the Pietists saw was the only safe way—toward the Holy Scripture itself. For even the *läsare* had been more concerned with reading books about the Bible than they were about reading the Bible itself, and Janssonism came to them with all the prestige of the archetypal Word. The penetrating insights of Luther, Arndt, and Nohrborg were thrust aside as though they were impediments on the true way.[5]

In the beginning of the movement, some of the *läsarepräster* thought it possible that Janssonism would bring a new and vital spirituality after the dry rationalism of the eighteenth century. Was it really as hard as Luther thought for God to create sinless men and women? Could not one rightly say to God, *omnia tibi possibilia* ("all things are possible with thee")? One could say what every pastor said, *justus et peccator simul* ("righteous and sinner at the same time"), but was it not more roundly biblical to say *omnia tibi possibilia?* After all, Jesus himself had said, "Ask and ye shall receive." And he spelled it out: "If you have faith as much as a grain of mustard seed, you could say to this sycamore tree, Be thou pulled up, and be thou planted in the sea, and it should obey you." Was it completely incredible that this peasant from Österunda, whose eyes flashed fire when he sold his pale flour, might have learned something about the bread of life which was hidden from the grave men of Lund and Uppsala? The wind blows where it listeth. Anyway, as the wife of a clergyman, Bertha Steinmetz, wrote to L. V. Henschen on March 27, 1848, it was hard "to find anything but good in the fact that Christians gathered together and encouraged each other with God's word." [6]

There were other reasons why counter measures were not instantly taken against the upstarts in Hälsingland. The unity of the Swedish church was being challenged from many sides during the middle of the nineteenth century. Two eighteenth-century movements had done especially well in Hälsingland—Pietism and Herrnhutism. The Pietists were strong around Bollnäs and Alfta, telling the farm people that it was not enough to be written down in the church records; one must be converted as well, and one must study Luther, Schartau, Nohrborg, and Arndt. There were Herrnhutists around Bergsjö, Delsbo, and Forsa, proclaiming Count Zinzendorf's *heart relation*, adoring the crucified Christ as though they were baroque Spaniards. Two other religious movements were not well known. Schartauism, with its grim appeal to the law and to the "order of faith," was rather closely restricted to the south and the west, and had made few inroads in Hälsingland. Neologism, appealing to intellectuals and urban types, was practically unknown in Hälsingland. But the consistory at Uppsala was not confronted by an isolated problem. A religious map of Sweden in the 1840s would show Separatists in Norrland, Baptists in southern Sweden, *Ny-läsare* ("New Readers") in Dalarna, *Gammal-läsare* ("Old Readers") in Hälsingland, Gästrikland, and Småland, and Methodists in Stockholm.[7] Among so many claims to special revelation, Janssonism could not at first have seemed to be a great threat.

There were also some of the clergy who, like Hans Norborg of Alfta, distrusted the Janssonists, but thought that the soundest strategy for the established church was to do nothing at all about it. One must not try to douse a fire with kerosene. Neither the primitive theology of this glib salesman nor his improvised polity and patterns of worship presaged a long life for these dissenters. The tactic of studied indifference had been used before by the Swedish church. In 1790, Mathias Stålber, pastor in Delsbo, had been threatened by *läsare* who were disrupting his discipline. He sent a memorial to the Cathedral Chapter at Uppsala, and urged them to invoke the seldom used 1726 law against conventicles. The chapter took a different line: "Religious liberty," they said to him, "as well as our own experience shows that at all times error is better dealt with by means of gentle conversation and persuasion, than it is by using the power of the law to terrorize the erring person." [8] In the same spirit, J. E. Tjerneld, pastor in Ovanåker, wrote to Archbishop Wingård that the best strategy was simply to ignore the Janssonists. Opposition, he thought, would inflame these enthusiasts, while passive tolerance would cause the movement to fade away.[9]

The newspapers of the time were widely critical of the clergy for what they took to be indecision and timidity, marks of the spiritual ineptness and lack of conviction which had permitted the Janssonists to spring up in the first place. In Gävle at the gateway to Hälsingland, *Norrlands-Posten* called for action against Eric Jansson as early as February 8, 1845, and frequent issues in the next two years repeated the call. *Aftonbladet* in Stockholm asked for action on February 8, 1845. P. A. Huldberg, who had no love at all for the clergy who had driven out Scott, used the columns of *Hudikswalls Veckoblad* to berate the establishment. On September 28, 1846, perhaps remembering some of the letters he had received from Jansson which contained so many misspelled words, he asked how such a helpless peasant could have developed so much gall. And he answered his own question: the clergy had left the door wide open through their worldiness. On August 2, 1846, *Söndgsbladet* in Stockholm said that the church was to blame for the emigration. *Göteborgs Handels-och Sjöfarts Tidning* agreed on May 7, 1849. One paper, however, *Thorgny* in Uppsala, reversed the more familiar reaction by pleading at first for strong action, on November 25, 1844, but changing its mind and saying on June 20, 1845, that after all toughness was unproductive, and that education in spiritual matters was a more profitable line. The change may have happened because the editor heard Jansson twist the Cathedral Chapter about his fingers in Uppsala and reported so on December 18, 1844.[10]

There were some of the clergy who were not vulnerable to the charge of weakness toward Janssonism. The opposition of N. A. Arenander of Österunda has already been noticed, and Risberg and Estenberg followed Arenander when their eyes were opened.[11] The wise and blunt Lars Landgren of Delsbo was known as a redoubtable foe of the Janssonists and expressed his gratitude for the Conventicle Edict. He was unimpressed by their logical coherence. "One needs only a half-hour of arguing in a talkative gathering of Janssonists or separatists, especially if women are talking, to ask yourself seriously if the devil has been set loose. Because if you ask them about a fence, they will answer you about a manse. [The pun is untranslatable: 'Ty så fort du frågar dem om gårdsgärden, så svara de om prästgården.'] When you talk about the right, they run on about the left. The weaker and narrower the mind is, the more it seems to be beside itself with happiness and pride over its extraordinary insights into the Lord's ways." [12] He reported to Baron Lagerheim that "at the beginning all the *läsare* were his followers [Eric Jansson's] but soon fell away by the hundreds, so that now only ten remain in the heretical sect." [13]

P. N. Lundqvist, factory pastor in Maråker, seemed to have answered Huldberg's charge that none of the clergy had publicly exposed Janssonism, because a few months later, in Gävle, he published the first full-dress account of the new sect and its beliefs. His book, *Erik-Jansismen i Helsingland* (1845),[14] is of primary importance because it was written by a man who was a contemporary and a close observer of the whole movement, and all subsequent studies have been heavily in his debt. His account is of course not objective. Lundqvist's avowed purpose is to expose the theological and biblical weaknesses of Janssonism, and also to defend the conduct of the *läsarepräster* in first befriending Jansson and then turning against him. According to Lundqvist, only the *läsarepräster* had been able to keep their flocks free of Janssonism.

But some of the clergy who were not *läsarepräster* were bitter about the role the clergy had played. A. A. Scherdin, pastor in Söderala, thought the churchmen were dupes for these Separatists, and fought a running battle with Jonas and Olof Olsson for their absence from common worship. Another implacable foe was J. E. Forsell in Torstuna. He, like Archbishop Henrik Reuterdahl, believed that no quarter should be given the Janssonists, and equally adament was N. S. von Koch, the minister of Justice, who laid the major blame for the sect at the door of the clergy.[15]

The reaction of the Lutheran clergy to Janssonism is incomprehensible unless it is understood in the context of the traditional Lutheran hostility to perfectionist motifs. Some crucial Lutheran doctrines must be thought of as *Kampfbegriffe*, positions taken up for strategic purposes in the battle which then was imminent, and this intolerance toward perfectionism must be so understood. Luther's foe at the time was Roman Catholic perfectionist illusions. He had himself been a monk, and knew at first hand the "counsels of perfection"—poverty, celibacy, and obedience. But he knew also that they had failed to bring him the expected peace of soul. Since they were available only to the special group called "religious," and also because of some inner dynamic that was so strangely corrupting, they soon stopped looking like virtues and began to take on the shape of pride, the first of the seven deadly sins. There was a progression in holiness, beginning with baptism; but there was no stage when perfection could be achieved. There was, to be sure, the perfection of Christ, which was pure; but the perfection of which any human being could achieve was impure. The worst sin a man could commit was to deny that he was a sinner.[16] If Luther had heard Jansson claim that he was without sin, Luther would have said that he was also without God.[17] No, the sorry truth is that as long as man walks the fields of earth he is *justus et peccator simul*, righteous and also sinful at the same time.

Monastic perfectionism was threatened by pride and Lutheran formulations made the most of the theme. The authoritative Augsburg Confession of 1530, epitomizing twenty-one essential Lutheran doctrines, was drawn up mainly by Melanchthon and bears the marks of his distinctive theology, but it was approved by Luther and was presented by him to Emperor Charles V. Article 12 was aimed at monastic illusions: "Christian perfection is this, to fear God sincerely, and again to conceive great faith, and to trust that for Christ's sake God is pacified towards us; to ask, and with certainty to look for, help from God in all our affairs, according to our calling; and meanwhile outwardly to do good works diligently and to attend to our calling."

No word here of sinlessness. The perfection hoped for is one of religious faith, the unwavering belief that God is pacified, and will not withhold his grace. Meanwhile, we are urged to produce some outward signs, doing good works. Melanchthon, according to Dilthey, wanted Lutheranism to develop along ethical lines, which would have brought Lutheranism very close to Wesleyan perfectionist ideals.[18] The phase "outwardly to do good works diligently" sounds like his pervasive moralism. His vision was that the culture of the ancient world and the intelligence of the Renaissance could be happily yoked together and baptized and transformed into a holiness ethically defined. But Lutheranism was to have a different history, making its distinctive contribution in the field of religion and theology, at the price of an ethical Quietism. If monastic perfectionism was threatened by pride, the Lutheran reaction was threatened by worldliness.

Luther believed that grace ought to bear fruit in genuine piety, and that daily life ought to be transformed by its presence; but the gift itself was in no way dependent upon these fruits. Perfection in Luther is a fullness of faith that one has been accepted by God, despite being unacceptable, and it has nothing to do with any monkish fulfillment of an ideal. There was no sharp line in his view between the town pastor and the town drunk, but on the contrary a strong bond between them: the bond of common sinfulness and the common need of grace.[19] The difference between them was that the pastor recognized his need and asked for help.

Hid from Jansson and from perhaps all Pietists is the Lutheran recognition that the Christian is a new man who is somehow the old man still, and that all of life shared this ambiguity. "In this," says Gustaf Wingren, "is the real meaning of *simul iustus et peccator.*" [20]

To Eric Jansson all this seemed like a terrible muddle, effectively concealing the Gospel version of salvation, which was deep, dazzling, and final. What could this passionate salesman from Österunda make of Luther's celebrated comment, "Sin boldly, but believe in Christ more boldly still"? To the Janssonists this was official approval of the man who made his annual visit to church Christmas morning with liquor on his breath. Lost on the believer in human perfection was the radiance of the Lutheran vision of Jesus Christ, so perfect and so pure that compared with him all human saintliness seemed corrupted, and in whose eyes no man living is justified. As Dietrich Bonhoeffer was to say, "For Luther 'sin boldly' could only be his very last refuge, the consolation for one whose attempts to follow Christ had taught him that he can never be sinless. . . . For before that grace we are always and in every circumstance sinners, but that grace seeks and justifies us, sinners though we are. Take courage, and confess your sins, says Luther." [21]

Lutheran doctrines of human sinfulness were drawn up to refute monastic perfectionism, but in the nineteenth century they were under attack from an entirely different quarter—sectarian Pietism. Another familiar Lutheran idea, the notion of the two swords, or the collusion between secular and sacred authorities, was brought into play to suppress the Anabaptists, but in the nineteenth century this idea was turned against the Janssonists. It was inevitable that when Eric Jansson assaulted the church on theological grounds he found himself arrayed against the state as well. It is no coincidence that when Comminister N. A. Arenander came looking for Jansson at Klockaregården, Österunda, in 1844, he included in the raiding party the officers of the crown.

The theological doctrine implicit is the sanctification of order in Rom. 13:1 ("Let every soul be subject unto the higher powers."). German Protestantism was committed to the Pauline idea that the Fall created an historical predicament in which man is continually threatened by demonic powers. Jansson agreed fully, but he would never have admitted that Pauline-Lutheran conclusion that the civil structures are ordained by God to protect mankind from chaos. The *imperium* is necessary, Luther said, to restrain evil men, and the *theologia imperii* was the grounds on which the state went into the business of restraining heresies. The Church of Sweden accepted the notion. When anyone in Sweden was born, baptized, married, or buried, the clergy noted the fact for the official records, and many civil rights depended on the clergy's certification of one's status in the church.

Defections from church authority were therefore punished by secular officials in a manner which is puzzling to anyone conditioned by the doctrine of the separation of church and state. The Wittenberg authorities were very harsh, punishing offenders against the church with fines, imprisonment for life, and sometimes execution. Compared with them, the Swedish administration of church laws in the nineteenth century was much more gentle, and punishment was meted out reluctantly. But all Lutherans agreed that heresy was a threat to social order and as such was punishable by the state. The doctrine led to hostility toward even such inoffensive ministries as that of George Scott in *Betlehemskyrkan*, and to violent wrath toward such dangerous heresies as those introduced by Eric Jansson. It is understandable that the victims of this wrath were not clear about its benevolent purpose, which was to protect Swedish citizens from the dangers not only of this life, but of the world to come.[22] The state, according to Luther, must do "the strange work of love," that is, the suppression of evildoers.

The theological object of Jansson's insurgence was the sinfulness of believers, and the political object was the unity of the Swedish people. The brothers Olaus and Laurentius Petri, who had studied under Luther himself, returned to Sweden in 1519 with the basic doctrines intact. At the great meeting in Uppsala in 1593, the controlling authorities of Swedish Lutheranism were established: the Scriptures of the Old and New Testament, the three classical creeds, and the Augsburg Confession of 1530. There was to be one people and one faith. Gustavus Adolphus had in mind a general consistory which would rule the country, made up of six spiritual and six temporal officers. The proposal failed to carry, though the state never denied its religious obligations. In Jansson's time the visible reminder of this unity was the minister of state for the Ecclesiastical Department.

The example of this unity which led to Jansson's undoing was the Conventicle Edict of January 12, 1726. In the early eighteenth century, various passionate evangelists were arriving from the Continent inflamed with messages which threatened to disrupt the ideal of one faith for all the people. The Conventicle Edict forbade all meetings in homes for religious purposes except those held by a single family for devotional purposes, or those in which the local pastor himself presided. The penalties set up indicated the seriousness of the concern. The first offense called for a warning. If there were a second offense, the fine was two hundred *riksdalers*, and for a third offense, four hundred *riksdalers*, or a prison term on bread and water for fourteen days. If anyone were so rash as to try it again, the penalty was exile for two years. A fine was levied even for listening to a minister from another communion.

Though the law looked grim on the books, it had actually been honored more in the breach than in the observance. Allan Sandewall has pointed out that in the whole of Upper Norrland, where conventicles were common, only four judicial proceedings leading to convictions were held in accordance with these regulations.[23] The new constitution of 1809, section 16, expressed the view which was much more in accord with the general Swedish attitude, and certainly was closer to the spirit of Enlightenment: "The King shall not force anyone's conscience, nor allow anyone else to do so, but shall protect everyone so that he may practise his religion freely, so long as he does not disturb the peace of the community or cause a public nuisance." When Charles ascended the throne in 1810, he had come from a Roman Catholic country and never really learned to speak Swedish. He was very sympathetic to the cause of religious freedom and refused to approve the infrequent court decisions against dissenters.

All that changed quickly with the rise of the *Ny-läsare*, Separatists, who objected loudly to what they considered the "works theology" in the new catechism of 1810, and to the hymnals of 1811 and 1819. When various ecstatic revivals broke out such as those which produced the "preaching sickness" of Småland in the early 1840s, and when Eric Jansson began challenging the hegemony of the state church, the Conventicle Edict was dusted off as an instrument for controlling disorder. The American Baptist temperance orator Robert Baird returned from a trip to Sweden in 1851 and was shocked at the punitive legislation. "A Mr. Nillson," he retorted, "has lately been sentenced to banishment, because he has become a Baptist." [24] He did not know that the purpose of the discipline was benevolent. Another temperance orator, his friend Peter Wieselgren, later Dean of Gothenburg, argued that strong legal measures were required to protect the sons of Sweden from those who would destroy the religion of the land. The function of the consistories, he said, should be to make sure that "the Swedish folk grew up in the Swedish state religion." [25]

Such was the character of the Lutheran church which Eric Jansson called an instrument of the devil. His blunt language was probably not so damaging as the review of the church in Sweden presented by George Scott after he had been driven out of the country: "If orthodoxy be vital religion, if uniformity be a church's unity, then the Swedish Church furnishes an almost unequalled example of religious unity; if comprehensiveness to the embracing of a whole population, and the effectual prevention of separate and in some respects rival denominations be the best condition of a church for developing the principles and practises of the Gospel, we may look for a flourishing state of things in Sweden. If large authority given to clergy, and efficiently sustained by the secular arm—if outward sacraments and observances—if legislative enactments, vigilantly watched over—can make men Christian indeed—then the Swedes are all Christians.—We shall see.—" [26] How Eric Jansson managed the inevitable conflict between Wesleyan perfectionism and Lutheran realism must now be our theme.

The Burning of the Books

> You insist on reading godless books, like those written by the damned Luther and the devilish Arndt.—But listen, you people.—Get this.—You're not reading God's word, but the devil's word!
>
> —Eric Jansson, *Nordiska Kyrkotidning* (1845)

The year 1844 began auspiciously for Eric Jansson. He sold his house in Sånkarby, Österunda, for nine hundred *riksdalers*, and moved with his family to northern Hälsingland. At first he rented two rooms at Kälkebo, south of Forsa, on the west side of Långsjön, and thought for a while of buying a home in Delsbo, where he had many friends. But this was not to be. However arrogant people like Risberg thought him to be, he conceived of himself as a helpless instrument in God's hands. He wrote in a letter, "When I was in Delsbo and planning to buy a home there to please my many friends, the Holy Spirit called to me and said, 'You shall go where I send you. You shall move to Forsa.' " [1] And so he moved—but not exactly to Forsa. He bought Lumnäs, at Stenbo, some four miles to the south. The owner of Stenbo was a distinguished local figure, Jon Olsson, a Forsa juryman. His son, Olof, had become an ardent Janssonist and had introduced his father to the prophet.[2] Olof moved south to Klockaregården, Eric Jansson's old home at Österunda.

Stenbo became his base of operations. He held meetings in farmhouses at Trogsta, Åkre, and Hamre. The movement was gathering momentum and was the chief topic of conversation in the homes of northern Hälsingland. There was some opposition, too. Karin Jonsson, from Utnäs, moved through the area holding rival meetings during which she denounced the new teaching. And the crowds who came to Jansson's meetings were not all friendly. At one meeting a man was particularly effective as a heckler when he quoted several passages from the Bible which stated clearly that forgiveness of sins is needed for believers. "You are so full of devils," cried Jansson, "that if they crept out of you the whole house would be full!" [3] At a gathering in May, 1844, after he had argued passionately that the sinner could be forgiven once and for all, a young girl challenged him: "But how then should I understand the passage from Job," she asked, "where it says that Job, under strong temptation, opened his mouth and cursed the day he was born? Wasn't that sin? But still God says of him, 'My servant, Job.' "

Jansson began fumbling through the pages of his Bible: "Hmmm, let's see, let's see." But after a futile search he said to her, "I'll talk to you about this some other time." [4]

It occurred to Jansson's restless and flamboyant mind that he needed some kind of dramatic event which would vividly demonstrate the strength of the Janssonist movement and at the same time provide a symbol of his followers' hostility to the words of learned men. He hit on the idea of burning all the books he could find except the Bible. An instinctive dramatist, and a shrewd analyst of the kind of people who followed him, he knew that elaborate theological arguments fall on deaf ears, while the sight of a flaming pyre of learning would stir his people to a frenzy. On March 19, 1844, he wrote a letter to his followers in Hälsingland proposing the holocaust. The Holy Spirit had visited him, he said, and had asked him "to fulfill in our time what is written in Revelations 18. . . . All books must be destroyed which are contrary to the Bible—and that means all the books I have seen." [5] Eric Jansson's most radical simplification was about to occur, uniting his following, as he expected, but uniting the opposition as well.

The motive was a hatred not only of theology but of learning in general. In the long and troubled history of religion there have been on occasions outbreaks of a strident anti-intellectualism. Francis of Assisi, for example, thought that the lyric grace of his faith was threatened by the learning of the schools, and he warned the Little Brothers against all books except the Bible and the missal. The Protestant sects have left a curious record in history, on occasion supporting learning and founding universities, and on other occasions expressing an *odium theologicum*. John Wesley, in so many ways close to the Janssonist position, aspired to be *homo unius libri*, a reader of the Bible alone; but his instincts were sometimes sounder than his avowed conviction, and he con-

Sophia Carolina Schön (1821–?) was born in Österunda Parish, Västmanland, and as a young girl became an ardent Janssonist missionary and book burner. After marrying a man named Peterson, she returned to Sweden in 1868. Painting by Olof Krans, courtesy of the Bishop Hill Heritage Association

tinued to read widely and to enjoy Shakespeare.[6] Early in the nineteenth century in Sweden, Eric Stålberg rejected all books of theology except the Bible.

Eric Jansson made no effort to conceal his hatred of learning in general. He had only minimal education himself, and despised the Lutheran clergy, who were all university graduates either of Lund or of Uppsala. He felt himself directly inspired by the Holy Spirit, while the learned clergy contented themselves with empty forms and impure theology, leading to spiritual death. In one of his hymns, he asks what was meant when Isa. 60:3 speaks of darkness covering the land. "Isaiah meant a flood of learning with its uncleanness which has drowned the people." One of Jansson's processional hymns, sounds the call to battle.

> The world's power is as nothing
> Compared with God's, who fights for us,
> Though learned men should watch
> As they have done throughout all time.
>
> When light has pierced the darkness
> The lesson of the past will show
> That fools will all be put to shame
> Though they should praise each other.[7]

Eric Jansson will be the instrument by which the Tower of Babel will be leveled, because he has been taught not by men, but by God himself.

> I have been taught by God,
> Who always stands beside me,
> While all His wisdom's light
> Shines through the heart's house.[8]

What should the true believer do in the face of this darkness flooding the land? Some of his followers answered, "Stop reading the books," but Jansson replied that this would not do. If they were to spare books which they did not want, their children would inherit this poison.[9] No, the Bible was clear enough, he thought. In Acts 19:13–19, he said, the Ephesian sorcerers burnt their books—though he did not mention that the sorcerers were confessing their own fraud and were burning their own books as a gesture of ritual purification. His more favorite biblical source was Revelations, chapter 18, in which Babylon, the home of demons, is seen to be beyond redemption, and the kings of the earth are committing fornication. Only flames will purify such dross. The particular dross Jansson had in mind was the books written by Luther and by Arndt. In *Nordiska Kyrkotidning* he explained why these books must be destroyed:

"The idols from which the heart must be purified are first of all the idolatrous books and leaders, especially Luther and Arndt. I can prove before kings and princes that Luther and Arndt wrote things contrary to the Bible. The wild beasts in the book of Revelation signify these false and devilish teachers, the great idol, Luther, and the murderer of souls, Arndt. What Luther says in one place, he tears down in another. In one place he calls his word God's word, his tongue God's tongue; but in another place he calls his own speech 'a chattering of words.' [10] He also says that he who presents himself as being sent by God, should prove it by signs. But this is false, since Christ says, 'An evil and adulterous generation seeketh after a sign; and there shall be no sign given to it.' (Matt. 12:39). Furthermore, Luther has given the Bible the coarsest of names, when he calls it a heretic's book." [11]

Per Niklas Lundquist, the schoolmaster and factory pastor in Maråker, quizzed Jansson carefully about this in preparation for the book he was writing on Janssonism. Jansson was more moderate to him. He admitted, "Luther was good for this time. But in his writings there is much that simply will not do for us. He calls his own writing 'a clatter of words,' and wishes them nine ells underground if they should prevent people from reading the Bible. Luther was only a human being. In the Large Catechism there are errors, though I can't yet say if they are his errors or his translator's." [12] Lundquist also asked Olof Olsson why the Janssonists condemned Luther, whereupon this conversation followed:

OLOF OLSSON: What have we to do with the dead? Christ lives!

LUNDQUIST: But not everyone has the gift to understand the Bible.

OLOF OLSSON: That's a strange thing to say. Have not all of his children received the Holy Spirit?

LUNDQUIST: Yes.

OLOF OLSSON: And can not the Holy Spirit understand what he himself has written?[13]

Eric Jansson taught such an identity with the Holy Spirit that there was no longer a problem of interpretation. He had come upon one of the weaknesses of the *Gammal-läsare* (Pietists who did *not* want to leave the church) which they themselves had begun to notice: an excessive interest in Luther and an inadequate knowledge of the Bible. J. Rolin, a moderate *Ny-läsare* (Pietists who did want to leave the church), associate minister in Hassela, expressed this in a letter to Lars Vilhelm Henschen on April 11, 1845: "It is curious that one party [the *Gammel-läsare*] . . . venerate Luther almost like a God, for instance in Wester and Norrbotten—yes, in Norway as well. They model their lives after Luther, rather than copying our Saviour. The other party, the Janssonists, condemn Luther. Both go to extremes. Nothing of this will lead to sobriety, missionary zeal, and Christianity." [14]

In the letter to *Nordiska Kyrkotidning*, Jansson goes on to say why Arndt also, the favorite of the pietists, should be burned. "Arndt speaks much about the spirit of truth, but this is the teaching of the devil, because when the Spirit of Christ is present, there is peace and joy." [15] Worst of all, Arndt accepted the Lutheran teaching on the ambiguity of human existence, stressing that even though one has been saved, he is still capable of sin and needs daily repentance. Perfection for Arndt was a goal rather than an achievement: "Though you cannot love perfectly, as God's word demands and as you would like to do, you should want to live perfectly." [16] Jansson thought such rubbish deserved flames.

The third satanic figure which Jansson proposed to burn was Anders Nohrborg. In the *Nordiska Kyrkotidning* letter he says this: "Nohrborg confesses himself whose spiritual child he is. 'I lie,' he says, 'like a wicked man on the rock of Christ.' But what does Christ say about such people on the last day? 'Depart from me, ye cursed.' (Matt. 25:41). God's word has been wasted from generation to generation. No one has become holy listening to talk like this. If you believe what I say, you will be holy, but if you doubt me, you doubt God himself. One man set out against me and my teaching, and how did it go with him? In three days he was jerked into infinity! . . . If you do not believe the message of the pure Gospel, which I preach to you, God will pour his wrath over you, and press you down into hell!"

It is not hard to see how this fustian would frighten a credulous audience, but it is a little harder to see exactly what he criticized about Nohrborg. He was an obscure assistant pastor in the Finnish church in Stockholm, who died in 1767 at the early age of forty-two. His sermons were published after his death and were much valued by the *läsare.* It is true that on his deathbed he said the words which so offended Jansson, but they were simply the Reformers' battle cry of *sole fide:* "As one who has no works of his own to show I lie on the rock of Christ. The enemies of my soul can shoot as many arrows as they are able: I hold fast, I hold fast to Christ, I shall not be overcome." [17] Perhaps the real ground of Jansson's animosity against Norhborg is revealed in his conversation with Lundquist: "Nohrborg," he said, "presents a long order of sanctification which is not in the Bible." [18]

But a bonfire needs to be stoked not so much with precise ideas as with explosive ideas, and Jansson had hit on a vivid simplification which would work on his followers powerfully. Luther, Arndt, and Nohrborg must burn, and with them, the books in which their ideas were inextricably mixed—sermons by Lindroth and Ekman and Petterson. "In all the books," according to one of the book burners, the papermaker J. E. Lundquist, "there was the false teaching that a man cannot be completely free of sin until he dies." [19]

Jansson picked the site of the first burning carefully. It was to be Alfta, where Anders Fraenell, the beloved pastor, had died on April 13. His successor, Anders Oldberg, did not arrive until two years later, in 1846, and the associate minister, Olof Ericsson, was the clergyman in charge. Alfta was the center of Jansson's strength in Hälsingland. So he passed the word around that there would be a great fire at Anders Olsson's house, Fischragården, at Tranberg, on the shore of Viksjö at noon on June 11, 1844. A large crowd of people from Alfta, Söderala, Ovanåker, and Bollnäs arrived for the event, *läsare* and others, some just curious and some carrying sacks of books. Boats arrived on Viksjön, and more books were added to the pyre.

One man tried to warn them of the consequences of such an act, but he was shouted down. Olof Olsson, from Kingsta, called out in his booming voice that if one book were to be removed from the fire, blood would flow! [20] Some economical zealots wondered if it would not be wise to save the leather bindings, and burn up only the paper, but Eric Jansson shouted, "You will be damned if you save the bindings from idols!" He seemed possessed, striding up and down exultantly: "See how Satan stares!" he cried, pointing to the crackling flames. "Satan had a feast to celebrate when Luther's works were published. Now he'll have to have a feast to drown his sorrow when they are destroyed!" [21]

The crowd stayed on and the bonfire turned into a meeting. Revelations, chapter 18, was read as a biblical warrant. Two farm boys cried out, "Thanks and praise to God!" and the crowd answered, "God be thanked and praised!" Per Andersson from Forsa, flown with excitement, cried, "We shall burn the legs off our prophets and priests!" [22] Olof Olsson spoke of passion and commitment, and said that he had given up house, home, wife, and children to preach the new faith. The parish constable of Bollnäs, Olof Bäck, proudly reported that he had burnt up twenty *riksdalers* worth of idolatrous books.[23]

In the evening Jansson went off to sleep at Olper in nearby Långhedsby. Next day Sheriff Holmdahl came to arrest Jansson, but was unable to do so, because Jansson was surrounded by a ring of ferocious followers. He came back the next day with a posse of forty men who had been drafted to help him. A soldier from the South Hälsingland regiment was with the party and described the battle that followed. The officers of the law moved in from several sides; there were terrified shouts from the women, blows were struck, blood flowed, and shots were fired. It was, he reported with astonishment, "a religious war in miniature." [24] Jansson was hiding in the attic, and the chimney had to be broken to get to him. But Jansson stepped through a hole in the ceiling and was arrested as he landed on the floor beneath. Were it not for the presence of a military man (perhaps the same person who wrote the letter to the newspaper) it was said that lives would certainly have been lost.[25] As he was being taken away to prison in Gävle, Jansson, who seems to have believed in the value of signs more than he had earlier reported, called over his shoulder, "If I come back you will know that I am sent by God!" [26]

The chances of violence breaking out in connection with Janssonist activity were very much greater from this point on. The formula was explosive: a passionate group of followers led by one with messianic presumptions, confronting the massive power of the Lutheran church and the state itself. The Janssonists from now on were marked by desperate partisanship and violent opposition. The sect did not itself willingly use violence, though it is true that Olof Ericsson of Alfta said to Pastor Lundquist just before the Alfta arrest, "It would be great if we had two hundred men. Then they couldn't arrest him." [27] In Alfta alone there were soon more than three hundred followers, but there was never any serious question of the power of the state to arrest the Janssonists. They were the victims of assault, and they sometimes resisted arrest foolishly, but they did not initiate violence.

At the sheriff's court in Gävle, Eric Jansson was given a preliminary trial, after which he was remanded to the custody of Västerås, which was nearer his home in Österunda. Here he was given an examination by a medical doctor, who was asked to determine his psychic health, and by a court chaplain, who was asked to determine his spiritual status. The medical man found him "an ecstatic, bordering on partial mental disturbance." The clergyman found him "suffering from two major and growing delusions, which inflate his sense of self-righteousness: the one stemming from his doctrine of sinlessness, and the other from his alleged call to preach the Gospel." [28] After being warned never again to visit the province of Gävleborg, he was dismissed, but the warning fell on deaf ears. He traveled about freely, reading the "History of the Passion of Eric Jansson"—an account of his suffering after his arrest at Alfta.

The old king, Charles XIV, had done nothing to hinder the Janssonists, but he had done nothing to help them, either, and during his last illness they prayed that he would soon be allowed to leave this earth.[29] When he died, in the spring of 1844, they made plans at once to travel to the capital in the hopes of seeing the new monarch, Oscar I. Their hopes were high. The young king was reputed to be far more liberal and tolerant than his father, and his first official acts were in keeping with his reputation. He dismissed Kristoffer Heurlin, the conservative minister for Ecclesiastical Affairs, and replaced him with a genial and well-meaning advisor, Frederic Otto Silfverstolpe, who had been secretary of the House of Nobles (*Riddarhuset*).[30] A new day seemed dawning for the Janssonists.

According to Jansson, he was sympathetically received by the king. Jansson was probably correct when he reported to his followers that Oscar I had no great affection for the ordained clergy, but it is hard to believe his report that the king "did not value the

Lutheran priesthood higher than the lice on a cow." The rest of the conversation was variously interpreted. Jansson said that the king asked him if he was seeking justice or mercy, and Jansson replied that he wanted simple justice. Thus far the accounts agree. But the newspaper *Stora Kopparbergs Läns Tidning* gives a different color to the conversation. According to that paper, when Jansson said he wanted justice, and not mercy, the king said rather crossly, "In that case, you had better take the matter before a law court. Goodbye." [31] Whatever the truth, Jansson lost no time in getting back to his followers in Alfta and saying, "I have visisted the king and won a victory." [32]

Nevertheless, his critics thought that Janssonism had begun to wane. Their hopes subsided when the news drifted through Hälsingland that Eric Jansson was back in Gävleborg Province, despite the warning of the court in Västerås, and was holding large meetings. A great crowd gathered at Jonas Olsson's house in Ina, near Söderala, on St. Michael's Day (September 28). Some of the listeners were believers, some were simply curious, and some came sceptically and remained to believe. After this meeting, he met with his followers in Forsa, and then went south to Bollnäs, while officialdom seemed paralyzed.

The adulation of the crowds inevitably developed his self-confidence, and he began to make extravagant claims for his role. On September 29 he said, "As certain as God is God the Holy Spirit talks through me. I have written a hymn with fifteen verses in an hour and a half. I can go to God in the darkness and say, 'Give me what I need,' while the learned men—teachers and bishops—have to sit and scratch their heads for every line." [33] But there are suggestions that he knew how thin the ice was on which he was skating. A correspondent who signs himself "B" heard him speak during this period and wrote a critical report for *Aftonbladet.* One minute Jansson bragged about the protection he had from the establishment, and in the next minute he said he was threatened by them. "It is time," this correspondent thought, "for the secular power to step in decisively. . . . Pastoral advice and warning accomplishes nothing with people as confused as these people are." [34]

The secular power was not listening, and on October 28, Eric Jansson staged his second book-burning, this time at Lynäs, south of Bergvik, on the shores of Lake Bergviken. This time even the psalmbook and catechism were not spared, Maja Stina herself throwing them into the fire and Jansson assuring his followers that he would soon replace them with his own books. After the bonfire a service of praise was held at Anders Jonsson's house, but this time the people seemed subdued, even depressed.[35] A hostile crowd gathered outside, and stones crashed through the windows where the Janssonists were worshipping. Olof Olsson and the massive Nils Hellbom rushed out and managed to catch two boys, whom they held hostage for the rest of the night. The sheriff arrived and Jansson's followers made their usual protective ring around him. There were angry shouts, some scuffling, and some clothes torn, but Jansson called out to his followers not to resist, and again he was placed under arrest.

Officialdom could not make up its mind about what to do with a man who broke the Conventicle Law, said libelous things against the clergy, broke the Sabbath, disturbed the peace, and at the same time claimed to be a harmless peasant who wanted to worship God in his own way and who had the blessing of the king. He was arrested, released, and rearrested. At the end of November he was in Gävle, undergoing tests for monomania. But on the twenty-seventh, the Cathedral Chapter at Uppsala sent word that he should be released. Archbishop Wingård said that he was really much less dangerous than the Almquist and the Ignell coteries.[36]

Early in November, immediately after the Lynäs arrest, Olof and his brother Jonas Olsson set out for Uppsala and Stockholm to marshal support for their cause. They went first to Stockholm to call on Carl Olof Rosenius, the friend and successor of George Scott, whom Jonas had known from previous visits to the capital city. Rosenius was the leading lay reader in the city at the time, being editor of the widely read *Pietisten* and also *Missions-Tidningen.* Though never a Methodist, but rather an evangelical Lutheran bent on reforming the church from within, he had been Scott's assistant at *Betlehemskyrkan*, and had been his devoted follower ever since he had heard Scott's lectures on the book of Romans. His theology was centered on the Bible, and he was very sympathetic to the *Läsare* movement, having in himself a strain both of Herrnhutism and Norrland Pietism. Let us hear his own account of the visit of the Olsson brothers which he wrote in a letter to Peter Brandell.

> Four tall, worthy *bedersbönder* ["upper-class farmers"] came in about four o'clock. They had just come to the city, asked me if I had time to talk to them, and gave me their names and addresses. (They are from Norrala.) The names Olof Olsson and Jon Olsson told me at once what kind of people they were, namely Eric Jansson's leading disciples, the chiefs of the bookburning party, come as deputies to ask the King for mercy and for freedom for their head, Eric Jansson, who, as you have probably read in the papers, has been arrested again and is in Gävle prison. I naturally thought this a rare opportunity and agreed to talk to them. Now began a conversation which I wish you could have overheard. It lasted three hours, until the clock struck seven, and I had to leave to attend a meeting at N's. Two of the men went along.[37]

Rosenius's report of the conversation which followed must be given in its entirety, because it shows the reaction of a sensitive and intelligent lay leader who was as determined as the Jansonists to be faithful to the Bible, and yet who came to a radically different position reading it.

> How did they defend their position? With the sloppiest and most astonishing talk, wrenching verses from their context. . . . When I presented the clearest biblical word and example, they flew off to another verse. If I spoke, for example, of some saint's sin and obsession they replied, "Yes, then they were fallen from grace."—When I asked, "If all evil in you is washed away, and all flesh dead, and you have no struggle, then you do not belong to the church militant, to Christ's soldiers, why then do you need the armor that Ephesians 6 speaks about?" They answered, "Yes we do. We still have the world against us, that is unbelieving men and the devil." But when I pointed out that the devil causes sinfulness, which you deny exists in a Christian, they contradicted themselves so badly that they said a devil cannot attack a Christian. "The devil," they said, "is laid out and bound." So they even deny that they have an enemy, which they had just said they fought against. I said, "But you have just said that the devil is the one we must fight against in Ephesians 6. The devil dared attack even Christ himself, even Adam and Eve, who had no sin, and Job, who accord-

> ing to God's own testimony had no equal in the land for piety, and Paul according to 2 Corinthians 12:7, etc." They changed the subject.
>
> It is impossible to count all their contradictions. I want simply to tell you of their manner of answering. Anyway, this is an instructive example of how far one can go in this line. Think of it, they sincerely believe that everyone who before Eric Jansson interpreted Scripture and learned Luther's, Arndt's and other teaching have been "the devil's apostles," as they themselves said. And now for the first time this little group of Christians has become Christ's only true flock! [38]

From Olof and Jonas's point of view, the visit was not a success, though Rosenius seemed to have been pleased with it. Some forty years later, when Jonas Olsson was the leading figure at Bishop Hill, he scoffed at Rosenius's report of the encounter. To begin with, he said, there were not four men who arrived, but only two: himself and his brother Olof. And their errand was not to beg for mercy; they were out looking for converts to their cause. Rosenius was said to be a Christian leader who had raised money for missions, and the Janssonists were obviously going to have heavy legal costs. Perhaps he would come to their defense. While they were talking, said Jonas, Rosenius called to his wife, who came in to meet them. These bluff, plainspoken farmers from Hälsingland thought that she was wearing jewels that a queen would have been proud to wear. So much for Rosenius as a supporter of missions! Finally, they were not asked to go to any meeting. In fact, there was no meeting at all, and no mention of one! [39]

When two honest men differ, an eavesdropper must wonder which is the true account, though the fact that Jonas Olsson was remembering an event forty years later must make Rosenius's story seem somewhat more plausible. We are certain, however, that Janssonism and what later became *Fosterlands Stiftelsen* showed irreconcilable differences. Rosenius recognized the danger of Janssonism and similar fanaticisms. A few days after the talk with the Olsson brothers, on November 12, 1844, he wrote to an old friend in Storkågeträsk, "At times I worry so much about all this that I long to die." [40] He remained convinced of the problematic human state, in which sin, the law, and death have been in principle overcome, but are still with us until we die. He warned the readers of *Pietisten:* "Never believe that you shall ever be free of sin in this life, or so pious that you will not feel like a deceived and sleepy hypocrite. Never believe that the law will not trouble you in your conscience, and judge you and punish you, or that death and the devil will ever fail to terrify you. No, you must suffer all this until the corruptible puts on incorruption. Then at last you may know yourself wholly free and unassailable." [41]

Jonas and Olof Olsson had much better luck at the end of November, when they called at the home of Lars Vilhelm Henschen in Uppsala. Henschen had been since 1835 an alderman in Uppsala and was known as a cultivated, devout lay leader who was anxious to lead a legal attack on the Conventicle Edict. He was a complex man, whose history had made him far more hospitable to Jansonnism than Rosenius could ever be. The son of a Lutheran pastor, he was from boyhood an admirer of J. Holmbergson of Lund, but he was also active in the Uppsala *läsare* circle. In his spititual genealogy there was a Moravian strain, together with some eighteenth-century Pietism and not a little of Schartauism. For a time at least he had been swayed by Swedenborgian romanticism, which was strong in Ångermanland, where he had been a judge for two years. He was firmly convinced of free congregational ideals which had been brought to Sweden from Anglo-American sources, and he was a friend of both Scott and Baird. Finally, he was a liberal, with ancestry in the eighteenth-century Enlightenment, and he liked to talk about such values as equality, liberty, and the natural rights of man. The effect of this complex mixture was that the Conventicle Edict offended him greatly, and he thought it out of step with the liberal constitution of 1809. The Olsson brothers had picked their man wisely.

If the question is put badly, "Was Henschen a Janssonist?" the answer must be, "Almost," or "For a little while." He wrote to his sister Augusta that he was impressed with the Olsson brothers' understanding, consideration, and forethought. He liked their "soft, sweet and yet bold speech, their friendliness towards those who persecuted them and their sympathy for those who must suffer for conscience sake." [42] His daughter said that for a few days he actually was a Janssonist.[43] But at least in the fall of 1849, he was anxious to refute such rumors and to dissociate himself from the Janssonists, saying that he had not been in 1844 a member of the party. He completely disowned later developments in the doctrine.[44]

His motive in listening to them and helping them prepare a defense was primarily that he thought them an excellent test case for the battle against the Conventicle Edict. He knew that they would suffer for their conscientious objection, and he promised that he would help them so that they would not suffer legally. But in the main he wished, like Thomander, to strike a blow for religious freedom in Sweden, and so he set about to prepare a statement of their case which would minimize their defections from pure Lutheran doctrine, and make them seem like dutiful citizens with conventional views which they could present plausibly and with restraint. When the Olssons came to town he had long conversations with them, together with C. A. Forsell, the senior assistant master in a school in Gävle, who was in Uppsala for canonical examination and ordination.[45]

The Olsson brothers had been charged with absence from church and refusal of confession by A. A. Scherdin, the pastor in Söderala. They were scheduled to appear before the Uppsala Cathedral Chapter to answer the charges on December 11, 1844. Six clergy sat on each side of the long table, and the bishop sat at the head, while the two brothers read the defense which Henschen had prepared. It was a document without a hint of revolutionary disturbance. The purpose of the Janssonists, they said, was to recover pure evangelical doctrine. They had studied the decisions of the Uppsala meeting of March 20, 1593, as well as the oldest, correct, and unchanged Augsburg Confession's twenty-one articles together with the attached decisions, and they could find in these documents no difference between the official faith and their own. As regards their doctrine of sinlessness, the matter was much misunderstood. Their position was summed up by St. Paul's words, "not I, But Christ liveth in me." *Flesh* in the biblical sense continues and is sinful, but it is buried with Christ. They did not deny that the converted man still has faults and imperfections which through daily renewal may be lessened, but they did not call these faults sins. How about the burning of books? Yes, they had been present at such burning and had helped burn them, but only, may it be noted, in concurrence with the wishes of the owners of the books. They had actually saved their own copies of Luther, in order to be able to refute the errors which they found there, such as his claim that original sin survives after conversion. They disliked, if the truth be told, certain words in the fifth prayer of the Lord's prayer in the *Smaller Catechism* ("forgive us our sins"), and also his account of the meaning of baptism by water. But perhaps the disagreement was more a matter of words than of substance.[46]

The consistory was completely carried away by the same charm which had captivated Henschen. Thure Annerstedt gave it as his considered opinion that the Janssonists were not heretical in their teaching on the ground and meaning of sanctification, and the chapter voted agreement. The brothers were solemnly warned not to take part in any activities which would divide congregations, and they were warned that they would lose their citizenship if they disobeyed the church. In return, the Olsson brothers promised to receive Holy Communion in the Söderala church, and then humbly made their exit. *Thorgny* reported on December 19 that the chapter had been unable to refute the Janssonist arguments. "Their confession was approved," wrote Lars Vilhelm Henschen to C. E. Ljungberg, knowing very well that he had written the confession himself.[47]

While all this was going on, Jansson was languishing in the hospital in Gävle, undergoing tests for sanity. A reporter from *Gefleborgs Läns Tidning* called on him and gave an interesting account of the visit. Jansson was writing a hymn when the reporter called. He expected to find an interesting psychological case, and was surprised to find Jansson an ordinary man. He said that Jansson's eyes had no fanatical gleam. His remarks were coherent and commonplace. The only exceptional quality he had was the ability to pour out dozens of biblical verses. Jansson agreed that not all men were called to expound Scripture. Quizzed about sinlessness, he admitted that "evil thoughts and temptations could come to him, but he did not call these sin."

"My final conclusion about this man," said the reporter, "is that he is neither mad nor is he a deceiver, but rather that he is a self-deceived *svärmare* [enthusiast]." According to this reporter the only cure possible now is "instruction by a teacher grounded in Christianity, with more than usual dialectical ability and knowledge of the Bible. If that should fail, there is no alternative but to keep him in jail, or send him into exile. It is of no use to hale him before the Consistory. He will answer anything with Bible verses, and if someone counters with another interpretation, he says that his view is the right one. [48]

On December 7, the same day on which Jansson was being interviewed at Gävle, his followers were staging their third book-burning near Jansson's home on the shores of Långsjön, at Stenbo, south of Forsa. In the absence of their leader, the event was planned by Jonas Olsson, Anders Olsson, Mårten Mårtenson, and the papermaker Jonas Nylund.[49] They chose a point called Runnudden, near the Stenbo farm, and most of the people arrived on skis. Sophia Schön, servant of Jon Olsson, and later a prominent figure at Bishop Hill, was an enthusiastic participant. This time they burnt all books, including some Bibles which were written in foreign tongues. The sheriff, Jonas Magnus Äström, arrived during the burning, and managed to pull some of the books from the fire, one with the symbolic title, *Sveriges rikes lag* [*The laws of Sweden*], published in 1734. A scuffle occurred.[50] In the trial which followed, Jonas Olsson was fined fifty *riksdalers* and was dismissed from his county position as juryman.[51]

The Cathedral Chapter now summoned Eric Jansson from prison in Gävle, and on December 18, he appeared before the consistory which had heard the Olsson brothers a week earlier. He repeated the central doctrine of his group, sinlessness, and defended his position with a pastiche of quotations from the Bible. There is no evidence that Henschen helped him with his defense, although it followed the main line established by Jonas and Olof Olsson.[52] He was released after being warned against proseletyzing, but returned to Lynäs and immediately planned a gathering of his followers at Anders Jonsson's home. On December 22, he was arrested there by Sheriff E. A. Flygt, together with fifteen of his followers, and charged with resisting arrest and breaking the Sabbath. He was imprisoned in Gävle, and held there for four months, until April, 1845.

There were signs that Janssonism fascinated the Swedish people, though they had trouble understanding it. These incredible farmers had put together an unlikely combination of piety and revolution, of moral pretension and criminality, of Bible study, prayer, and insolence. Many observers from Archbishop Wingård on down tried to sort out the creative elements in the movement—if any such existed—from what seemed like demonic forces loosed on the gentle Swedish landscape. The interest and apprehension of the people was sensitively analyzed by Emilie Flyggare Carlén in her novel, *En natt vid Bullarsjön* (1847).[53] There were signs already of the disruption which was about to occur, but for the moment they caused only vague uneasiness and an intuitive judgment that this wild-eyed Eric Jansson and his wheat flour customers boded no good for the land.

Mounting Fury

You insect, you can't understand the Bible!

—The Reverend Lars Landgren to Eric Jansson

After their mild rebuke from the consistory at Uppsala, Jonas and Olof Olsson returned to their homes in Söderala and blithely continued holding conventicles. Pastor Scherdin was incensed both at them and at the Cathedral Chapter, which, he thought, had glossed over the differences between orthodox Lutheranism and the Janssonism which he had encountered. Annerstedt's friendly attitude toward Janssonism at Uppsala was beyond belief. In any case, it was he, their pastor in Söderala, who had denied the Olsson brothers Holy Communion, and it was he who was not about to readmit them to the altar without stronger assurances of their orthodoxy. He summoned the two of them and three other Janssonists to a meeting of the local church council at Söderala on January 26, 1845.[1]

"Why have you left the church?" demanded Scherdin.

"Most of us have been faithful churchgoers," replied Olof, "until you began using humiliating words. Before the whole congregation you spoke of Eric Jansson as 'a traveling tramp,' and those who support him as 'deluded followers.' "

He said that Pastor Scherdin and his associate had talked in this vein Sunday after Sunday, but neither one had made any effort to correct the Janssonists by presenting biblical evidence.

"What then are your beliefs?"

"We think we have Abraham's faith," said the voice of Olof, though the hand was the hand of Henschen, "not according to the teaching of men, but according to the Bible, and we want without being stopped by men or by laws to find true holiness in Jesus Christ."

Janssonism, they insisted, was consistent with the theology of the Uppsala meeting of 1593 and with the Augsburg Confession.[2]

But when Scherdin pressed them, they admitted that the report of their faith given to the Cathedral Chapter was misleading, and that they had very serious criticisms of Lutheranism. They renounced the confession they had given to Annerstedt. Jauntily they said they would not have had meetings of their own during church time if they had not been banned from attending the regular service. Now they could promise not to burn any more books, since all their books had been burned! Scherdin reported all this to his superiors at Uppsala, and said that so far from being penitent and orthodox, the Janssonists remained heretical in their teaching and contemptuous of the church.

The Cathedral Chapter called them back to Uppsala on April 30. Not in the least intimidated, when the Olsson brothers passed through Gävle on the way south, where their leader Eric Jansson was in prison, they stopped and held a public meeting. They were immediately arrested but were released when they said that they were on their way to Uppsala to stand trial. At Uppsala the solemn clergyman were just as bland as they had been in December at their first encounter. The Olsson brothers promised to attend services in the church, if Pastor Scherdin would permit them to come. They accepted another solemn warning and were excused.[3] Doubtless they called on Henschen, who at that time was busily trying to help Maja Stina Larsdotter get her husband Eric Jansson out of jail. The charges against him, said Henschen, were absurd: he is "now in jail, now freed, now called insane, now sane." [4]

On February 17, 1845, Klockaregården, Jansson's parental home in Österunda, was sold to Olof Stenberg, "Olle in Stenby," as he was called. The letter of contract, which is still kept at Klockaregården, was signed by Eric's brother Karl (who was given power of attorney for his imprisoned brother), by his other brothers Pehr and Jan, by his sister Anna, and by his mother, the widow Sara Ersdotter. According to the terms of the sale, Eric's mother was to be allowed to stay at Klockaregården and would be given a small share of the produce of the farm.

Pastor Arenander meanwhile had not rested in his battle against the sectarians. He brought to trial at the Torstuna District Court Anders Pehrsson from Domta and Pehr Jansson from Norrbångsbo.[5] The charges: they had failed to attend church services (until 1855 punishable by law). Their defense was prepared

Olof Stenberg (1818–92), who moved from Stenbo, Forsa Parish, to Österunda, followed the Janssonists to America, and became one of the colony trustees. He later became a Methodist minister. Painting by Olof Krans, courtesy of the Bishop Hill Heritage Association

by Henschen. They had not attended church, they pleaded, because of scorn they felt for the worldly lives lived by the clergy. Besides, they pointed out, they had been forbidden to receive Holy Communion. How can anyone be banished from church and at the same time he punished for being absent? They were given severe penalties, but the supreme court modified their sentences after Henschen appealed. On March 17 the Svea Court of Appeals in Stockholm also overthrew the Eric Jansson arrest order, saying that the evidence presented was not strong enough to warrant further incarceration. Eric Jansson was free to return home in April. "Poor province!" cried the editor of *Norrlands-Posten*. "Poor local authorities! Perhaps poor Eric Jansson!" [6]

The day after Pentecost, on May 12, 1845, a group of Janssonists gathered at the home of Anders Pehrsson at Domta. Present were Olle from Stenbo [7] and Per Andersson of Torstuna, who had shouted, "We shall burn the legs of the clergy" at the Stenbo book-burning.[8] Lars Quick of Delsbo was also there, as was the omnipresent Anna Maria Stråle, the servant at Klockaregården. Threats had been made against them, and they knew the growing danger of their meetings. They wondered whether it was wise to hold meetings in the face of so much hostility, until the fiery Per Andersson said, "Shall we draw back because we are weak in the flesh and are afraid of mere men?" Somewhat nervously they started to sing a hymn. A crowd began gathering outside the house, and they locked the doors. But Erik Strömberg, juryman in Alfta, wrenched a window loose from its frame and forced his way in. He was accompanied by Anders Andersson, the churchwarden from Bångsbo, angry because his four daughters had become *läsare* and were presumably at this meeting.

The sheriff, J. E. Ekblom, responded quickly when he heard of the fracas, but he was too late to stop the assault. Anders Pehrsson claimed that he had been beaten, but Ekbolm could see no visible marks of this. Olof Jansson showed him the bruises he had received. But the most seriously hurt was Olle from Stenbo. They had dragged him from the house by pulling his hair and throat, and then had banged his head against the wall. Blood was pouring from his scalp, causing Wilhelmina Westerberg, a tidy housekeeper to the last, to run with a bowl to catch the dripping blood. They reported the assailants as saying to Olle, "You seven thousand barrels of shit, before you yanked at the reins of a cream-colored horse, now see what you can do with a brown horse." [9] The women tried to prevent them from tying him into the driver's seat of a cart. Two of them ran for the sheriff, whose arrival put a stop to the carnage. But no one has brought to trial for the offense.

Further indignities were heaped upon the Janssonists. At one meeting in June, crowds broke up the service and the Janssonists hid in a barn. According to *Hudikswalls Veckoblad*, Maja Stina Jansson hid in the loft but slipped through the hole in the floor and hung suspended by her skirt, to the delight of the attackers, who amused themselves by beating her with branches as she dangled above them. [10] People known to be sympathetic to Eric Jansson were verbally abused and sometimes physically attacked, while the authorities looked the other way.

On Midsummer Day, June 24, 1845, many people from neighboring towns gathered at Jonas Olsson's lovely farm at Stenbo. People sat entranced on the grass in the courtyard between three farm buildings, while Jansson sat on a platform with the Bible in front of him and expounded the Scriptures. The local sheriff, J. M. Åström, had assembled a great many voluntary deputies to assist him, and he had asked each of them to carry a club. They ordered Jansson to stop preaching, and when Åström was climbing the stairs to arrest Jansson, a woman grabbed the sheriff by the legs and pulled him to the ground. During the scuffle that followed, Jansson escaped. Åström thought that he had hid in the house, and in order to search for him broke down the door to the house. [11] But Jansson, accompanied by Olof Stenberg, raced through the woods, and together they made their escape from Runnudden by boat. They rowed to the southern tip of the lake and then started off through the woods. At Enånger they stopped to watch people dancing about a maypole, and at Norrala they saw the same thing. At dawn they arrived at Ina, near Sörderala, where they received food and drink from Jonas Olsson. [12] Sheriff Åström followed Jansson there and tried to break up a meeting which was being held. During the fighting at that time one of the Janssonists wielded a knife and cut a man from his mouth to his ear. [13]

The rising crescendo of violence seemed intolerable not only to the authorities responsible for law and order but to all the people of Hälsingland. It drove the Janssonists to despair. Olof Olsson and his wife made plans during the summer to go to America and look for a place for the Janssonists to settle. On Saturday, July 18, he and his wife Anna Maria started out from their home at Kingsta and planned to visit relatives at Järsvö, in order to say good-bye before leaving, perhaps forever. When Helena Jonsson was an old lady, she remembered what happened that Saturday afternoon at her home in Hamre, even though she was only twelve years old at the time.

The time was noon of a very warm day, and the Olssons stopped at the Jonsson farmyard in order to rest their horses and also to say good-bye to their Janssonist friends. Jon Jonsson was pleased to have one of the sect leaders as his guest and thought that other Janssonists in the area should be invited to hear him speak and tell about his America dream. A friendly crowd soon began to assemble, but an even larger crowd of ruffians also gathered. "If you will all be quiet," said Jon Jonsson "you are all invited to stay, but if you are not quiet I beg of you to go away, since we plan to hold a worship service." They grabbed him roughly and threw him out of his home. Olof Olsson protested that he was simply stopping briefly on his way to Järvsö, where he was going to say good-bye to relatives before leaving Sweden for good.

"We'll help you on your way!" they shouted, and tied him to the seat of the wagon.

"Such helpfulness is more than I had a right to expect," said Olof. As the horse left the farmyard with its bound driver, he managed to turn his head and call back to his friends, "We shall meet again across the water—under a free flag!"

The next day, with the Olssons safely on their way to Järsvö, the little band of Janssonists at Hamre debated about whether to capitulate to this opposition, or whether to go right on worshipping as they pleased. They hit on the compromise of meeting secretly on the shore of nearby Lake Ingan. Undiscovered, they held their meeting and listened to Mårten Mårtenson read the Scripture and expound the Sermon on the Mount. But the crowd of hostile neighbors again assembled back at Jon Jonsson's house. They found no one home except the old grandmother, who was told to be quiet while they searched the house to see if Olof Olsson had come back. One of the searchers, a nephew of the grandmother called "Tolfman Olof" from Bole, smashed a mirror into bits, and in the process cut his finger. He came downstairs with blood streaming from his finger and told his laughing comrades, "Look! Auntie bit me on the finger!" Fifty years later, sitting in her home at Bishop Hill, Helena Jonsson was still angry. "Was it strange" she wrote, "that after such events and such inhuman treatment, the thought came to us and later the decision was made to seek a free country somewhere where God's Word would not be bound?" [14]

As though anxious to make up for their role as worried spectators during 1844, the establishment by the middle of summer, 1845, was in full cry against the Janssonists, determined to use the strong arm of the law to silence them once and for all. On Saturday night, August 16, the household at Klockaregården, next to the Österunda church, was quietly sleeping. The house now belonged to Olof Jonsson, but he was away. In one room slept the widow Sara Ersdotter, Eric's mother, and in the center room slept the two Janssonist girls, Anna Maria Stråle and Sophia Schön. At 11 P.M. there was a sharp knock at the window and the girls sat up in bed. A deep male voice said, "Is the swindler at home?" [15]

"No swindler lives here," said Sophia.

"Yes, he does—a tall swindler from Hälsingland named Olof Jonsson."

"He is not here."

"I don't believe you, because you are swindlers. Let me through the door or I'll break through a window."

It was Pastor Arenander together with a posse of angry men. Sophia was not easily frightened. She called out to them, "Do you have a sheriff's warrant?"

"We shit on all sheriffs," someone replied. They broke open the window and entered the room where the two girls lay with the blanket under their chins. Pastor Arenander asked Sophia what she was doing in Västmanland when she lived in Hälsingland. She told him that she had once lived here, had relatives here still, had a pass from Pastor Schilling to come here, and was helping with the haying and the grain harvest.

"Get up," Arendander said to the girls. Embarrassed to be seen in their nightgowns by a roomful of men, they refused, whereupon Pastor Arenander grabbed Sophia by the hair and pulled her out of the bed to the floor. Not giving the girls time to dress, he made them get into his wagon and drove the three and one-half miles to Torstuna, where Sheriff Ekblom lived.

Bouncing along the valley road to Torstuna, Sophia could not resist heckling her captor. When he asked her why she was a Jansonist, she replied that she did nothing of her own will but simply was an instrument of her Father's will. She said that Arenander sprang at her, yanked her hair again, and shouted, "That's God's will!" She told him that violence done on the highway was especially punishable under Swedish law, and delivered one more dart that may have found its mark: "Are you," she asked, "planning to tell this story when you preach tomorrow?" Ekblom listened to her story, interviewed all the other participants, and carefully wrote down what they said.[16]

The reason Eric Jansson could not be found was that he was hiding under the straw in a cattle barn on the Torstuna road. Anna Maria Stråle remembered that after the Stenbo scuffle, a price of thirty crowns had been put on Eric Jansson's head, and he lay hidden for five weeks.[17] Anna and Jansson's wife, Maja Stina, brought him food from Klockaregården, traveling through the woods at night the long way around to avoid detection. One day Sara Maria Persdotter, the wife of Eric Blom, the soldier, came to Anna Maria and asked her if she knew where Jansson was hidden. "I can't tell you," said Anna Maria, "but follow me if you want to see him. If I give you a sign, it will be all right for you to see him." She talked to Eric Jansson and received his permission to let Mrs. Blom come to the barn. When Mrs. Blom saw how pale and thin he looked, hiding under the hay, she fainted.

It seemed to these women that he would be discovered if he stayed longer in this hiding place, so they moved him to the Blom home in Källsta, Österunda Parish, where he was hidden in the attic. One day Blom's son, Eric, heard a noise and asked, "What's up there in the attic, mother?"

"Rats," his mother replied. But it was clear that Jansson could not hide safely there for very long. At midnight, Anna Maria went with Eric and Maja Stina to Uppsala to find Henschen. Henschen wrote a petition to the king which they took to Stockholm. On the way they met A. M. Ljungberg, who had once been a sailor, and who now peddled Bibles around Uppsala while being supported by the American Seamen's Friend Society.[18] He gave them tips on how best to approach the king.

The king was not at the palace when they got there, and when he finally returned, they were unable to get an audience with him. One day they heard that the queen's mother and her lady-in-waiting were walking in Djurgården. They went there and had a promise from her that the king would see them the next day. But the servants turned them away. Luckily, they managed to talk to the king as he was stepping into his carriage, and he told them to tell K. W. Björkenstam that they should be allowed to be tried in court. They traveled home.[19]

It was clear by this time that the local authorities had mounted a determined campaign to rid themselves of Janssonism, which was disrupting the peace, ruining the worship of the church, and now threatening to take hundreds of men, women, and children away from their homes to an uncertain future in America, breaking up homes and families in the process. But the highest authorities were still unable or unwilling to move. N. S. Koch on August 5 warned J. E. Söderlund that no action should be taken against Jonas or Olof Olsson at this time.[20] Nevertheless, Jonas and Olof were brought to trial at Norrala on August 20, and both confessed to having broken the Conventicle Edict. Olof announced that he intended to go to America. "Since you will not receive or believe the truth," he told the court, "I shall have to go to the heathen world." [21] He was found guilty and assessed the usual fine. Henschen used the case to test the Conventicle Edict, and appealed the decision to the supreme court, which upheld the decision in May 1846.[22] But by that time Olof and his wife and two of his children were safely in America.

During the summer, Jansson missionaries had found their way into nearly Dalarna and had won a number of converts there. Among them was Anders Blomberg, the tailor, who was brought before the court at Mora on September 8. When he was reproached for using brandy and other strong drinks, his defense was that he had been misled by the Lutheran clergy. Unused at this time to the kind of eccentricity which was familiar in Hälsingland, the puzzled court at Mora decided that he was the victim of partial mental derangement which had taken a religious character.[23]

Jansson's catechism and hymnbook had by this time achieved some circulation and was thought to have been a prime cause for the growth of the movement. Consequently, the courts moved against the Jansson printer, C. G. Blombergssom, who had published the books in his small printshop at his home in Ina, Söderala. Since he had printed "Söderhamn" on the title pages of both books, he found himself vulnerable to the charge of having published under false pretenses, and he was forbidden to print anything more. On September 27, 1845, his shop was taken out of his hands and given to Lars Bergroth, who moved the presses to Grafverbergska, just outside Söderhamn, and began printing the newspaper *Helsi* on February 5, 1847.[24]

The Janssonists were trying to present a low profile, but they refused to give up their meetings. On October 18, Eric Jansson gathered a few of his trusted friends at the home of Matts Källman, a charcoal burner who lived at Överbo, near the Voxna factory. Jonas Olsson was there, and so was the printer, C. G. Blombergsson, still smoldering over the loss of his press. There were other people from Alfta, Söderala, and Ovanåker. While they were at prayer, Källman's daughter Margareta looked out the window and cried, "The yard is full of people!" It was what had become by now the familiar assault, and they knew what defensive measures to take.

Eric Jansson was quickly hidden in a special cell under the floor. Erik Källman went out the back door as a decoy and ran through the snow into the woods, pursued by several men who fired shots at his speeding figure. Jonas Olsson ran in the other direction hoping that he would fool them into thinking he was Eric Jansson. The ruse worked, and several men followed his tracks in the snow. When they finally caught up with him, Jonas kept his left hand in his overcoat pocket, pretending that he was Jansson, who was known to have lost two fingers on his left hand. They finally wrenched his hand free, and when they discovered that they had caught a decoy, they banged his head on the floor so hard that he had trouble afterward putting on his hat. They found Peter Källman, the other son, hiding under a bed, and they banged his head on the floor so hard that many years later, when he was sixty-eight years old and living in Bishop Hill, he could still show the scars on his head.[25] There was no recourse in the law to find redress for such outrageous attacks, since the Janssonists had lost their citizenship when they left the Lutheran communion.

Jansson's final trial in Sweden was planned for Delsbo at the end of October.[26] Jansson decided to submit voluntarily. On his way there from Gävle, he stopped at the home of Olof Johnsson at Valla, Söderala.[27] As usual, they planned a meeting for the evening, and the word went around the Janssonist circle to assemble at the Johnsson home. But Pastor Scherdin heard the news as well, and dispatched the constable, Olof Pehrsson, to arrest Jansson. The constable came into the house, grabbed Eric Jansson by the collar, and cried, "Now, Eric Jansson, you are mine!"

"What is the charge against me?" asked Jansson. "Am I a thief, or a rapist, or a murderer?"

"I have been sent here by Pastor Scherdin," said Pehrsson, "to place you under arrest and to take you to the prison at Gävle."

Eric Jansson protested that he had just come from Gävle, that he had a pass to travel, and was in fact on his way to a trial in Delsbo. Suddenly someone turned the lamps off, and in the darkness the huge Janssonist, Nils Hellbom, grabbed Eric and said, "Follow me!" They went out the door and the hostile crowd let them pass, thinking that one of the officers had Jansson under arrest.

"We've been fooled!" shouted Pehrsson's men. One of them rushed out of doors, caught up with Jansson, and tackled him. Hellbom pulled the man off Jansson and told Eric to run. He then confronted the mob and told them the truth: Jansson was voluntarily giving himself up to be tried at Delsbo, and they were interfering with justice. Placated, they returned quietly to their homes. Jansson spent the night in the woods with a friend, "K.," and made his way next day to Enånger, where he arrived in the afternoon. He slept the night with his friend and supporter, Pehr Pehrsson, and next day arrived at Delsbo for trial.[28]

Archbishop Wingård spent the summer on the trial of the elusive Jansson but had to report very little success. Writing to a friend, Samuel Kamp, the archbishop said, "During the summer I pursued Eric Jansson, and since then I have been on a Fall trip, during which I scolded and warned him in the very parish in which he was born. A *läsare-präst*, bewitched by Jansson's gift of speech and his ingenuity, recommended him to his adherents in Hälsingland and allowed him to hold conventicles. These developed into open feuds with the church and the priesthood, and afterwards led to an emigration in order to hold the followers together. His power over them was demonic, and soon he put himself in the place of God and our Savior. Nothing so arrogant has ever been seen in our land. Surely if we had a stronger government this mischief would not have been able to befuddle and defy the civil order. It is incredible that he could thwart the inquiry and make his escape to America *in salvo*." [29]

But Jansson did not thwart the inquiry at Delsbo. They had trouble finding a judge who would sit for the case, but finally the court was convened at nine o'clock on the morning of October 30, with Judge C. G. Levin presiding. The room was full of spectators, most of whom were violently partisan, though not so visibly as the lady who took up her post prominently at the back wearing a Delsbo costume. She stood with rapt attention with hands folded when Jansson was on the stand, and turned her face to the wall when Pastor Landgren spoke.[30] Jansson began inauspiciously by telling the judge he was born in Biskopskulla, Västmanland (he meant Uppland). The trial began.

> JUDGE: You have held conventicles, Eric Jansson?
>
> JANSSON: Yes, I have.
>
> JUDGE: What right have you to do a thing like that?
>
> JANSSON: I must obey God rather than men. I serve a high priest who sits at the Father's right hand, who lives forever and who intercedes for us. If he were on earth he would not be a priest at all, but now he is in heaven and sits at the Father's right hand. He has sent me, through the calling of the Spirit, to proclaim his mercy to sinners and to offer them salvation.

Judge Levin then turned to Pastor Landgren and asked him what passages in the Bible would be relevant to these views. Landgren replied, "I do not know."

> JUDGE: (to the clergyman who was to soon become Bishop of Härnösand) You who are a clergyman do not know? (Turning to Eric Jansson) Do you know?
>
> JANSSON: Hebrews 7:24–26; Matt. 11:27; Matt. 10:19–20; Luke 14:17–24.

But Landgren was soon to appear at better advantage. He spoke at some length about his difficulties with Janssonism. The Conventicle Edict may be a pagan law, he conceded, but it is in any case the law of the land; and it has also served a useful purpose, for example, to provide a restraint against such excesses as Janssonism. The attack which Eric Jansson launched was not aimed alone at the church, but against all civilization and culture. The same theme was developed by Pastor Schilling from Forsa. Sheriff J. M. Åstrom told about his problems in enforcing the law against the Janssonists, and he was supported by Soldier Schön, by Anders Lif, and by Jon Andersson. Erik Palme said that he had once seen Eric Jansson in church, but Jansson was sleeping and had to be nudged awake. Jansson denied that this charge was true.

A long letter written to Archbishop Wingård by Landgren describes the extreme tension of the time and the hatred which the Janssonists had engendered against the church officials: "Three farmers," Landgren wrote, "rushed to the sheriff and demanded that I be brought to court for having said that all they do is evil, and that they are heretical in their teaching on sanctification." [31] He felt the blame for the Janssonists should be placed on higher officials in church and state who had failed in their duty. We would never have had this problem, he said, "had the matter been dealt with on a higher level, as it should have been, and the whole weight of the madness not dropped on a single officer, who has to fight for his own life against the coarse bands of depraved people who flourish here. Wherever I turn I meet scenes which make me weep tears of blood.

"The anger of the people against me goes beyond all bounds, and it was lucky for us all and for Eric Jansson himself that he was put in jail, because had he not been I don't think the day would have ended well." Landgren had had his fill of Eric Jansson. "If the supreme court lets him go again," he wrote, "I hope he never comes back." He felt that the worst fate which can occur to a clergyman had been his: to have been rejected by his parishoners. "None of my advice or special prayers will govern this wild crowd," he wrote. "After the Janssonists went unpunished in Forsa, they think themselves authorized to take over in place of the appointed authorities, against whose weakness they noisily complain." [32]

The decision of the court was that Eric Jansson should be remanded to the custody of Gävle prison, and held pending further investigation. The case was continued until November 15.[33] Prison guard Sven Jacob Pira from Samtuna was instructed to escort Jansson to Gävle, and on Thursday noon, November 21, he led the prisoner through town on the road to Hudiksvall. Jansson said as he left that he would like to go to America.[34]

When the guard and his prisoner reached Hudiksvall, they rode south along the coast road toward Enånger. Pira proved to be very sympathetic, and may even have been a Janssonist, because as they rode along he confided a startling bit of information. "I know what they have decided to do with you," he said. "You will be kept in a cell called 'seven' [*sjuan*] together with another prisoner who has a life sentence. They have promised to let him go free if he can kill you in one way or another. Haven't you any friends who can help you?"[35]

"Yes, I have," replied Jansson. When they came to Hålsange, he wrote a note saying that the authorities planned to take his life, and managed to deliver it to Pehr Pehrsson, the *nämndeman* ("juryman") at whose home he had stayed a short time before on his way north to Delsbo. Pehr and three other Janssonists found out the road which would be taken by the prison guard, and set up an ambush. When the prison party came to Uppånge, Pira turned his prisoner over to another guard, and so did not need to be a knowing victim of the ambush. At Lanås crossroads, not far from the scene of the November book-burning, three men rushed upon the wagon, pulled the warden's coat over his head, cut the reins of the horse, and bound the warden with the harness. They took off through the woods with the kidnapped prisoner. The guard managed to work free from the reins, run to a nearby house, and tell Jonas-Martha of Färgebacken the whole story. Next day a woman who lived near the crossroads spilled goat's blood at the intersection and spread the story that Jansson must have been murdered. Willing to use a little deception in the service of the greater truth they were called on to proclaim, Maja Stina Jansson went about town in mourning clothes and sighed when anyone mentioned Eric Jansson's name.[36]

America Has It Better

> Amerika, du hast es besser
> Als unser Continent, das Alte.
>
> —Goethe

In October, 1845, Jonas and Olof Olsson made a trip to Stockholm which proved to be of decisive importance in the history of the Janssonists. On the way they stopped overnight by the roadside south of Uppsala and were alarmed to hear a man driving up in great haste and asking if Jonas and Olof Olsson were there. It was Henschen. He made arrangements to see them next day at Rotebro Inn, and when they met there, he gave them an address where they could meet in Stockholm. Henschen thought he was closing in on his quarry, the Conventicle Edict, because he now had in his possession what he took to be an effective test case.

But Henschen was still not absolutely sure in his own mind that his clients were orthodox in their teaching. He asked them if they had ever read *Concordia Pia,* written by the Pietist P. J. Spener in 1675, and when they replied that they had not, he went out and bought a copy and read selected passages to them. He asked them if they could accept the theology of these passages, and when they said they could, he suggested that they kneel and pray together. At the end of the prayer, the well known alderman from Uppsala asked the two farmers from Hälsingland if they would bless him. Together they went the next morning to the palace, where Cabinet Minister Silfverstolpe took them to see the king, Oscar I. "Haven't these people burnt books?" the king asked.

"Yes," said Silfverstolpe.

"Have they burnt the Bible as well?"

"Oh, no, Your Majesty. They love the Bible and use it for their rule of life."

"I shall see to their request and grant them their rights," said the king.

They went out elated at this success and even Silfverstolpe told them that if they ever needed his help, he would give it to them.[1] But as usual, their high hopes were dashed. It would take a long time before they received the same rights which drunks and evil-livers received—namely the chance to be heard in court and the privilege of being admitted to the sacrament. Lars Erik Bergroth from Skåle, Söderala, came down to warn them of the bad news from Uppsala: the Cathedral Chapter had finally lost patience and voted that "if either of these two brothers holds a meeting of *läsare*, directly or indirectly, they shall be instantly arrested and banished from the country." The word was passed that the king and the clergy were in complete agreement on this, and that if the Janssonists met again they would be sent into exile.[2]

Jonas was then tried in the local church at Söderala. Witnesses appeared who said that he tried to get them to leave the state church, but he denied this. One witness said that at the meetings they merely read the Lord's Prayer and some verses from the Bible. "Surely," said Judge Björk, "it is no crime to read the Lord's Prayer and some verses from the Bible!" Jonas was freed, but he was now disenchanted with Sweden. With great elation he sold his home at Ina and prepared to move his family to America. But the journey to the Promised Land was not to be that easy. It seemed to the harassed Janssonists that they were being, by turns, threatened with exile and also forbidden to leave Sweden.

In the chancellery office at Gävle he was denied his passport, Pastor Schilling having refused to sign his travel permit. He went immediately to Stockholm and consulted Henschen again, who drew up a very strong petition to the king. He also had the help of Carl David Arfwedson, the American consul in Stockholm.[3] When he took the revised petition to the palace, the usually urbane ecclesiastical minister began to roar like a lion, and Jonas Olsson thought that Siberia was in the offing. But he spoke with quiet dignity: "Believing the Minister's friendly promise of help on our previous visit, I have come to ask for your help and advice. We want passes to travel to America. We have nothing more to ask than simply to travel away from here and leave our fatherland in peace. That is all. The only solution we can see is that we abandon our struggle here. But the chancellery office is closed to us, and has denied us our passports."

King Oscar was true to his word. When they left Stockholm

The bark *Wilhelmina,* which carried 119 Janssonists to America in the fall of 1846. An elderly woman and 22 children died en route, and 3 babies were born. Courtesy of the Gävle Museum

they carried with them the royal permission for all Janssonists to leave Sweden by any port they chose. A few days later, on September 6, 1845, Olof Olsson received his passport in Gävle, and he and his wife and two children and a tanner named Erik Hellsten boarded the *Neptunus* and set sail for America.[4] He was to serve as a spy in the Promised Land, dispatched to find out exactly where the milk and honey flowed most freely.

Meanwhile, Eric Jansson was trying to make his escape from Pharoah's minions. After the Voxna Bruk battle, he fled on skis with Pehr Källman eastward to Sofåsen, Ovanåker, on the Alfa road; and here he lay concealed for seven weeks under a barn floor. Jonas Olsson and "Bröd-Jonas" Malmgren heard that he was about to be arrested, and took him by night to Sven Olsson's house at Grängsbo, Alfta. Unfortunately, Sven tended to be garrulous, especially when his tongue was loosened by brandy. On a trip to Gävle, he overheard some people in a tavern wondering aloud where Eric Jansson was hidden. "At my house," he volunteered. A woman who was a Janssonist heard about this breach of security and walked night and day the sixty miles from Gävle to Alfta to warn Jansson.

It was clear that further evasive action must be taken, and they decided that in Dalarna, just to the west, was the best hope of safety. His first stop there was with the Löfquist sisters at Mora, on the tip of Lake Siljan, and from there he went to Siljansfors, Johannesholm. Accompanied by Pehr Pehrsson, he then traveled down the road past Vibo and Vimo until he came to the home of Lindjo Gabriel Larsson in Malung. For the time being, at last, it looked as though he had found a secure refuge.

Dalarna was second only to Hälsingland in the hearts of the Janssonists. Stina Lisa Boman, from Garpenberg, was the first Janssonist missionary to that colorful province. Filled with youthful ardor, she had preached there for four days running, and had stopped only when she was arrested. She told the officials that she was sorry, and they sent her back to Hälsingland.[5] Two girls from Dalarna, Liss-Hans Maja and Stina Oldsdotter, had heard Jansson speak in nearby Hälsingland and had brought back the message to their neighbors. Several Janssonists had been assigned Dalarna as their territory, and one of them, Sophia Schön, was well known in Mora and Malung.[6] The other apostles to Dalarna were John Hansson of Grangsbo, Jon Persson of Älvkarhed, and Anders Olsson of Tranberg, each accompanied by an entourage of women. The missionaries had made a strong impression because of their appearance, the men wearing long hair, uncut and unbraided, which descended to their shoulders, and the women wearing identical fur collars and wide white hats.[7] Jonas Nylund, the papermaker from Delsbo, was credited with having converted the Bouvin sisters and a man named Kihlström.[8]

Jansson was hiding successfully from the local police, and it is difficult for us to pick up his trail with the spoor over a century old. It is likely that the chief place of his concealment was with the Gabrielssons in Östra Fors, Malung Parish, Kopparbergslän. In a story strikingly like that of the Stenbo family, Lars, the eldest son of the prosperous Lindjo Gabriel Larsson, had first heard Eric Jansson speak in Hälsingland and had gone back to Dalarna to tell his parents and neighbors about the new winds of the Spirit blowing nearby. Lindjo and his wife had six children. The Gabrielssons were the most influential converts in Dalarna, and were to play a prominent part in the early history of the colony. It is certain that with the Gabrielsson home as a base, Jansson made forays back into Hälsingland wearing various disguises. Otto Söderlund, pastor in Bollnäs, wrote in March 5, 1846, that Jansson "visits Bollnäs, Alfta, Ovanåker, travelling in a wagonload of hay, or covered market cart, sometimes dressed like a woman, and so pushes his work forward." [9] The work he had to do was to persuade twelve hundred followers to sell their earthly properties and move to America.

The ability of the Janssonists to endure suffering cannot be described without astonishment. The explanation must lie in the absolute certainty they felt as to the rightness of their cause. Saints expect to suffer in the devil's world, while the worldlings flourish like green bay trees. The stoicism of the Janssonists was not based on any profound belief in the natural law, nor was there any of the voluntary endurance of suffering which Catholic centuries had called asceticism. It was rather the calm acceptance of their role as soldiers in a country occupied by enemy troops. Jansson's hymns frequently use this theme, for example in Hymn No. 28:

Should we not suffer from the sword,
We never shall find peace,
That is the way to serve our Lord
In wicked times like these.

We all must suffer savage blows
From Satan's evil men,
For throughout time their rage will show,
Our Shepherd's regimen.[10]

But even heroic suffering for what seems a worthwhile cause must take its toll, and the strain upon Eric Jansson, physically and psychically, was very great. J. E. Ekblom saw him just before he left for what proved to be his America trip, and the alderman from Torstuna was shocked by what he saw. Ekblom had had a dream the night before that Jansson had become insane, and so he looked at him with special interest when they met. He thought he saw signs of mental disturbance. "Just before he travelled to America," wrote Ekblom, "he looked as though he should be taken to a madhouse. There are instances of insane people who, except for their fixed ideas, can discuss other subjects coherently and even with shrewdness." [11]

There is no sharply defined frontier of mental stability, and all we can say certainly is that during 1845 Eric Jansson was a deeply troubled man. The picture of himself which he projected to his following was that of the messianic savior of the world, appointed by God to be His agent for bringing in the Kingdom. It is not hard to find a label for this kind of fixation: we have before us a classic case of delusions of grandeur. Various psychic maladies are attendant upon this kind of megalomania, but if you add to them the fact that most of his countrymen did not think of him as a savior at all but simply as a deluded troublemaker who had to hide himself like a thief in order not to be locked away in Gävle, you have a recipe for a frightful tension, and it is not surprising that Eric Jansson showed traumatic symptoms.

Certainly as he skied across Dalarna, accompanied by Lars Gabrielsson, he felt the excitement and apprehension of pursuit, but mingled with it was the heady prospect of escape with a large body of followers and the creation of a city of believers in the country of the free. Reports drifted through Sweden that he had been seen wearing a woman's dress and cape, and there was one rumor that he had a baby in his arms, though this would seem to overdo the needs of camouflage and to make skiing difficult. It was said also that he had pulled out his two prominent front teeth in order to make his disguise more effective,[12] but this is surely untrue. During the previous fall he had reported very bad toothaches, sometimes making it difficult for him to speak, and when he was on trial at Forsa on October 11 it was noticeable that some teeth were missing from his mouth.[13] The teeth were pulled because they were infected.

Even his headlong flight had some gratifying moments. When he stayed at the home of his followers, he could see that the hostility of the state had fanned the flames of his friends' admiration, and he was treated with devotion as though he were divine. On what was to be his last night on Swedish soil, at Älvdalen, he slept in a room which was afterward made into a shrine. No one was ever allowed to sleep in the room again, and the bedspread he used was later sent to Bishop Hill and treasured as a relic there. Leaving Malung, on his way to Dalby, in Värmland, he spent the night at Femtryan, the home of a well-to-do plant manager, Jan Jansson. He and his wife and their seven children, the youngest only a few months old, became so enchanted with the fleeing prophet and the exciting prospects ahead that they sold all their possessions and joined other Janssonists in Christiania, shortly afterward sailing for America on the *Tricolor* with a large party from Dalarna.[14]

Something of the terror of Jansson's last days in Sweden, together with his idea of a miraculous deliverance, is conveyed in a letter which he wrote to L. V. Henschen on February 10, 1847, and which has been preserved among the Henschen letters in the Uppsala University Library. The stormy weather seemed to have been sent by God to make possible his escape: "I could see God's power, which forced my enemies to stay in their homes until I, like Jesus himself, could slip through their hands." He said that he had been given sanctuary by various friends, and had left the home of Lindjo Lars Gabrielsson in Malung on March 23. The office of the provincial governor in Gävle had written to the Norwegian officials and warned them that Eric Jansson would be arriving; but they had been mysteriously blinded, and he and his wife and children traveled unmolested "by water and by land to New York."

Before leaving Norway, Eric Jansson wrote a long letter as a parting word to his followers in Sweden. His curiously mixed mood is visible in the letter, which alternates between self-pity and delusions of grandeur, between sorrow at leaving his native land and anger at those who made him go. A snowstorm was swirling outside his window. The letter was written with a pencil, since he had not been able to carry pen and ink with him during his headlong flight, nor did his hosts seem to have any to lend him. He had suffered physically. The shoes he had worn did not fit him very well, and the ski trip through the mountains had been excruciating. Nor did he have overshoes. Lars Gabrielsson and he had quenched their thirst by drinking snow water, and in the process he had caught cold; but he had now recovered and was able to send one last greeting to his beloved friends.

He told them that there should be a fast during the last three days in April, presumably when he left Scandinavia. He then gave them a list of topics for their prayers. First, that all of them should meet that summer in America. Secondly, if there was anything in Eric Jansson that was a stumbling block to the faithful, they should pray that it be removed. He wanted very much to be truly perfect, and especially that God would enable him to speak with many tongues as he left his native land to take up residence on a foreign shore. Third, that all dry branches in their midst should be cut off and cast into the fire. Fourth, that their printer, C. G. Blombergsson, should be spared the suffering his enemies had planned for him. Fifth, that all hypocrites be cast out of the holy fellowship. Sixth, that God either open ways for the faithful to get to America, or else take them up to heaven. Seventh, that those bewildered followers (probably he meant Olof Olsson) who had been led astray by the seductions of Pastor Heström should be brought back into the fold. Eighth, that Sweden be made to suffer worse than those wicked cities Sodom and Gomorrah. Ninth, that if anyone doubts that these things will happen, he should be cut in two, or else made to drop dead like the lying Ananias and Sapphira.

They can imagine, he wrote, the feelings he had in this desolate spot when he thought about his imminent departure to a new land. There had been faults, of course, in his spoken words and writings; but one thing had not been at fault—his sense of the possibility of life. He asked that the letter be copied and sent around to all of his followers in Ovanåker, Bollnäs, Alfta, Söderala, and also to those who lived in Västmanland.[15]

The other document which survives from those last days before leaving for America is his *Afskedstal*, his farewell address to all of Sweden, especially, he said, to those "who have rejected me, even though I was sent by Jesus himself."[16] The charge was that he had broken away from lawful arrest and imprisonment. The truth is, he pointed out, that he had been given many chances to escape, none of which he had taken. He had even on one occasion submitted himself freely to the legal authorities, in order to answer charges brought by lying clergy. On that occasion he had been given no free trial at all, because his own witnesses had not been allowed to testify, and the prosecution witnesses had been told what to say by "the Delsbo pastor."[17]

Before leaving, he wanted everyone to understand that he had not despised the light which the state church had to give but was in rebellion only against those devilish clergy who were disloyal to that light. A certain strain of megalomania, never far from the surface in Jansson, broke out in his final greeting: "I cannot help telling you the truth, though you do not want to hear it, that the words I have spoken in the place of Christ shall judge you on the last day, since he who despises me has despised God himself." [18] He had never departed from the classic Lutheran formularies—the Uppsala Conference of 1593, and the twenty-one articles of the Augsburg Confession. He hoped that the eyes of the newly crowned monarch, Oscar I, would be opened to the devilish heresies of the clergy! So he bade farewell to the leaders of the land, who were soon to bite their tongues off when they saw the punishment coming their way. And farewell as well to all those poor souls who believed the clergy rather than his own message. He was leaving now, but he thought it likely that he would return to Sweden before he died, since no prophet ever died in any country other than his own. His heart was a turmoil of love and hatred, of fear and braggadocio, of tenderness and cruelty.

Eric's wife, Maja Stina, traveled separately to Norway in a party of seven with her two children, Eric, seven, and Mathilda, three; two women from Falun; another woman identified as a sister of Sven and Louis Larson, who later settled in Victoria; and Olof Norlund. The stratagem was that Eric and Maja Stina would use the passports of Mr. and Mrs. Eric Larson of Kingsta. The Larsons were Janssonists who had embarked on the ill-fated *Ceres*, which floundered off Öregrund, and after their rescue they had decided to return home. To keep the passport tally correct, Olof Norlund traveled with the party until they reached Christiania. He then turned over Mr. Larson's passport to Eric Jansson, and returned home.[19]

They decided not to risk arrest by waiting too long in Christiania, and so boarded the first outgoing vessel, which happened to be bound for Copenhagen. From there they proceeded by rail to Hamburg, where they boarded a ship for Hull, England. They crossed England by rail, and sailed from Liverpool across the Atlantic, reaching New York in late March or early April, after a voyage of six weeks. The movement from one means of conveyance to another in midwinter must have seemed nightmarish to Maja Stina, who was pregnant, and when they arrived in New York she delivered her baby stillborn.

For the next few years a steady stream of Janssonists crossed the Atlantic, each one suffering as much as their leader had. They were married and single, young and old, men and women—alike only in their belief in the American myth. The Delsbo *Husförhörsbok*, the pastor's house-examination record, lists during the 1840s long rows of local citizens who emigrated, the oldest one being eighty-six years old.[20] Jon Jonsson of Delsbo was seventy-seven years old, but that did not stop him from hobbling up the gangplank of the *Norrland* on crutches. The Janssonists aboard persuaded him to go back home, giving the local paper a chance to say, "Truthfully, he won out in that deal." [21] Another elderly person fell into disfavor with the Janssonists on the *Norrland*. She stood by the rail and sighed, "We know well enough how hard it has been up to now, but God only knows the trouble we shall have from now on." That pleasantry was enough to get her put ashore before the *Norrland* sailed from Hudikswall on June 20.[22]

Some Janssonists decided not to go to America, for example, Mårten Mårtenson, who stayed in Sweden and died in his native land in 1898. Eric Jansson's younger brother, Carl Lindewall, left his wife and started off with a party of Janssonists, whereupon his wife said, "If that's the way he feels, good riddance." But Carl turned back before boarding the ship and spent the rest of his life as an honored Swedish citizen in Domta, Österunda.[23] A man and his wife and children, who were persuaded that Jansson was a true prophet, sold all their property and came to Gävle to buy passage with eleven hundred *riksdalers* in their possession. But he met some old cronies in a tavern and proceeded to get drunk at a farewell party. The Janssonists refused to let him board the ship, and his wife and children left without him. His wife gave him three hundred *riksdalers*, kept eight hundred for herself, and gave him this parting advice: "Be concerned about your soul's salvation, as I am concerned for my own!" He sickened shortly afterward and died.[24]

Some of the clergy at first refused to sign the necessary papers to permit the Janssonists to leave Sweden. On May 11, Arenander wrote to the Cathedral Chapter at Uppsala that he would not sign the passport for Eric Olsson and Jon Andersson, hired hands at Klockaregården, and that he would refuse to sign for any other Janssonists.[25] In some cases, parents or husbands prevented a sailing. Six girls who arrived at Christiania with the early party were intercepted by their parents and taken back to Sweden, and one husband forced his wife to return.[26] Lars Jonsson, church warden from Stenbo, woke up one morning to find that his wife had slipped away at dawn to join a party of emigrating Janssonists. He followed her as quickly as he could to Hudikswall, only to find that she had already left on a sailboat for Gävle. Hurrying there, he found her in the company of some Janssonists, and simply asked, "Do you want to go home with me now, Kerstin?" She went back to Stenbo with him, and during a long, happily married life which followed, the incident was never mentioned again.[27]

Wild stories about the Janssonists began drifting through Hälsingland, among them the fable that a group of Ovanåker had sacrificed a lamb and a goat, and had danced about it with primitive glee.[28] The clergy warned their parishoners to guard their children, since there were rumors afloat that the Janssonists were planning to sacrifice a child. Just before the *Norrland* was ready to sail from Söderhamn, a woman came to the dock and screamed across the water that her eleven-year-old daughter had been kidnapped and forced to go with the Janssonists. She was invited to search the ship for the child, and did so but without finding the girl. Still certain that her daughter was aboard, she cried on the dock and was invited to search the ship again. This time she found the girl hidden under a pile of hay, and on the deck they had a tearful reunion before leaving for home.[29]

It was probable that some of the Janssonists travelled under false passports, such as Eric Jansson must have used to leave Norway. Lars Landgren wrote to the sheriff in Gävle on October 9, 1846: "I have heard that certain of Eric Jansson's sect have left their homes to go aboard some vessel in Gävle without proper permission, and have equipped themselves with false passports. This presumption is grounded on the fact that a farmer in Källeräng, Delsbo, whose name is Jan Johnsson, was found with such a fake passport made out for his wife. No passports have been given out by me except one to farmer Lars Persson in Oppsjö. All other passports which are alleged to be signed by me are false." [30]

Jonas Olofsson, Söderala, wrote a sad letter to the sheriff in Gävle. He said that his wife, Anna Magnusdotter, and his daughter, Margta, had a clergy pass to travel; but his daughter Karin, who was twelve years old, had no pass and was supposed to stay in Sweden with her father. She had an arm which was crippled from birth, and it was thought best that she remain in Sweden. But she had disappeared, and her father was sure that she planned to emigrate with her mother and sister. The sheriff was asked to search for Karin and if he found her to send her back to her father in Söderala.[31]

On May 29, 1846, the Janssonists stood poised for departure with the royal permission to leave from any port in Sweden (except for Eric Jansson, who was a fugitive from justice and was to have been tried by the courts). Rumors said that the Janssonists had assembled some fourteen thousand *riksdalers*, each one having contributed what he had to a common purse, and that they intended to buy a ship of their own to be under command of Master Mariner Lexelius.[32] But if this were part of the early planning, there is no evidence that it was ever carried out.

There were no passenger vessels plying between Sweden and America in 1845, and the Janssonists had to book passage on vessels built for carrying freight on which passenger accommodations could be improvised. The first of the Janssonists to leave was Olof Olsson, who left to reconnoiter in company with his wife, his two children, and a journeyman tanner named Erik Adolf Hellsten.[33] Olof wrote an enthusiastic letter from New York on December 31, 1845, in which he praised the reception he had received from the Methodist minister Olof Hedström, described in glowing terms what he had found, and promised his fellow Janssonists that "God's blessing awaits all of you in the future when you come here, in Jesus name, to set your feet on this new land." [34] One other Janssonist dared a winter crossing, P. W. Wirström, who sailed on the ship *Fanchon* at the end of 1845; but the main body of Janssonists left in a series of vessels beginning in the spring of 1846.

The first group to sail left on May on the *Ceres*, a freighter loaded with iron and carrying a few passengers. One of the sailors in the crew, Victor Witting, who later became a Methodist minister and served churches in Sweden, described in his memoirs the voyage of the ill-starred *Ceres*. There were sixteen or seventeen Janssonists aboard, including the vivacious Sophia Schön, Sven Larson, Jonas Malmgren ("Bröd-Jonas," the baker from Bollnäs), Margareta Malmgren, the wife of Jonas Erickson, her mother and father, and a number of young people. On Sunday afternoon, when the voyage had hardly begun, a violent storm broke out, threatening to capsize the *Ceres.* It was reported that one of the Janssonists, thinking that the end was coming for all of them, shouted to one of the crew that it was time to think of his salvation. The sailor replied, "We've got too damn much to do now to think about salvation!" [35]

Though the *Ceres* and all aboard seemed doomed, the Janssonists stood on deck and prayed and sang as though they were in a sunlit farmyard in Hälsingland. Victor Witting noticed their composure with astonishment. It was as though they really believed in a life after death and looked upon their imminent departure from this world not as a disaster but as a joyous entrance into a brighter world. Miraculously, the *Ceres* did not founder at all, but was washed ashore at Öregrund, its prow wedged in some rocks in such a way that all the passengers could be rescued and taken back to Gävle. Witting never forgot that nonchalance. In the summer of 1847, when he was in Stockholm looking for work, he met another band of Janssonists bound for America, applied for a job with the crew, and was signed on as a steward especially responsible for the Janssonists.[36]

The next ship to leave with Janssonists aboard was the brig *Tricolor,* which put out from Christiania in early June, 1846, carrying the party from Dalarna. Seven sturdy young men left for America from Forsbyarna, and on the way slept at the summer pasturage Råberget under a tree which is still called "the America pine." Among the Dalkarl party was the wealthy Lindjo Lars Gabrielsson, who sold all his extensive belongings and property and used the money to pay the passage of many impoverished companions. The *Tricolor* arrived in New York without mishap on July 25. The next ship to leave was not so fortunate. The schooner *Betty Catharina,* commanded by Capt. Anders Rönning, left Söderhamn on August 8 carrying sixty Janssonists, who were sharing the hold with a cargo of iron. The ship simply disappeared. It was sighted off Öresund on August 15, 1846, but it was never seen again.[37] Possibly the cargo of iron shifted in heavy seas causing the ship to capsize and sink without leaving a trace; but no one will ever know. One of the passengers who went down with the *Betty Catharina* was Erik, one of the sons of Lindjo Lars Gabrielsson.[38]

When the Norwegian brig *Patria* left Stockholm in early June, 1846, it had forty-four Janssonists on board under the direction of

Anders Andersson. But a larger group boarded one of the best ships in the Swedish merchant fleet, the brig *Charlotte*, which carried one hundred and fifty Janssonists when it cleared Stockholm in August.[39] Each had paid one hundred and fifty *riksdalers* for the crossing. The voyage was a series of crises, taxing the leadership abilities of Jonas Olsson. Two of the women were pregnant when they sailed, and the two babies were delivered before the ship reached Denmark. An old man died before they reached New York. The captain had more than navigation to occupy his attention: he presided at births, baptisms and read an appropriate passage from the Bible when bodies were dropped into the sea. He had a great deal of help in all this from Fru Hebbe.[40]

Peter Källman came alone on the *Vistula*, which docked at New York on September 19, 1846. Two days later the bark *Wilhelmina* arrived with one hundred and nineteen Janssonists aboard. Nils Hedin, the preaching tailor from Jämtland, had been in charge during a particularly difficult crossing. These Janssonists had more than their share of troubles, vandals having stoned the place in Gävle where they had been staying before they sailed. On the high seas, Captain Bock presided over the burials of twenty-two infants and one old lady. Three children had been born in mid-Atlantic, adding their tiny cries to many sounds of a ship at sea.[41]

The stately Norwegian brig *Agder* hove to in New York on September 28, 1846, with thirty-nine Janssonists aboard, among them Sophia Schön and the faithful papermaker, Johan Lundquist. But it also carried some passengers who looked askance at these religious fanatics, among them the Norwegian doctor, Gerhard S. C. H. Paoli, who was on his way to practice medicine in Chicago and also to help found the Women's Medical College in that city. He wrote a letter to his friend, Theodore Schytte, a medical student of the University of Christiania, and Schytte kept the letter to fill out a book he was writing to serve as a guide for emigrants. Dr. Paoli used the three months of the passage to study the behavior of these strange people. When he came on board in Stockholm, he found them all gathered for a service on deck, and he was struck with the look of elation on their faces. He saw no signs of sorrow when the gangplank was raised and they left their native land, presumably forever. When the *Agder* rounded the southern tip of Sweden, heavy seas made the ship pitch and roll, causing many of the Janssonists to become seasick.

As a medical doctor he was interested in hearing their leader deliver an impassioned sermon to a suffering congregation, the theme of which being that if anyone became seasick he must have lost his faith. During calmer days on the voyage the Janssonists tried to convert Paoli to their faith, plying him with Bible verses; but he remained incredulous, and was pleased when he saw that five of their number jumped ship when they docked at Helsingör.

On the Atlantic another fierce storm broke out which was only stilled, so they assured him, when they had invoked the aid of Eric Jansson. On sunny days he saw them stretched out on deck in small groups, reading the Bible, discussing the meaning of the verses, or singing one of their long hymns. They were confident that though they knew no English, they would be empowered by grace to speak the new language the moment they landed. Mischievously, Doctor Paoli had the American pilot address them in English when they came near land, but they only stared at him uncomprehendingly. Who is to blame for all this kind of religious imposture? he asked in his book on the immigrants. The clergy? The state? [42] There was no answer.

In August the brig *Caroline* lay beside the wharf at Stockholm, taking on supplies for an Atlantic voyage under the command of Captain Jacobsson. A reporter for *Söndagsbladet* strolled about watching the preparations and was impressed by how old the Janssonists seemed. Some of them, he said, had barely enough strength to ascend the gangplank, and he wondered how they possibly could be pioneers.[43] They had a chance to find out, since unusually rough voyage lay ahead of them. An American ship pulled into Le Havre on October 9 and reported having seen the *Caroline* with its foremast broken off. A letter from Captain Jacobsson reached Sweden on October 30, stating that his vessel had been washed ashore near the island of St. Pierre, in western Newfoundland. He had managed to transfer all his passengers to another vessel bound for Boston, and they had all finally arrived safely at New York.[44]

Three more vessels left in the fall of 1846, the *Solide* and the *Norrland* from Gävle, and the *Fritz*, out of Stockholm, the latter having 117 Janssonists aboard. The *Solide* under the command of Captain Danberg, reached New York in three weeks. No other vessel left with colonists until October of the following year, when the brig *Sophia* sailed from Gävle with 95 people from

Bollnäs aboard. The brig *Edla* left Stockholm shortly afterward with 2 Janssonists aboard, and the bark *Augusta* made its way into port at March 8, 1847, after thirteen weeks at sea. Then came the bark *New York*, which left Gävle on October 17, 1846, and reached New York a half year later, in March, 1847.[45] The *Elize* had a faster voyage, leaving Stockholm in July and arriving in New York on August 26, 1847, carrying 28 Janssonists.

There were no Janssonists at all crossing the Atlantic in 1848. The ship *Cobden* sailed from Gävle at the end of June, 1849, carrying a large party of Lutherans under the leadership of L. P. Esbjörn. Ten Janssonists rode along. The brig *Pehr* arrived in New York on October 10, 1849, with 15 Janssonists, and on September 15, 1850, the bark *Aeolus* arrived with 144 Janssonists, led by Olof Stenberg, who was making his second trip.[46] The bark *Condor* out of Gävle, loaded with iron and carrying 107 Janssonists, came into port on November 9, 1850, under the command of Captain Starr, and the *Primus* arrived in New York on June 20 of that year. Only one other large group of Janssonists was to come, this time on a vessel which landed in 1854.[47]

The departure of some twelve hundred men, women, and children—the first mass emigration from Sweden—caused a great deal of anguished questioning on the part of those who remained behind. Sweden asked itself what had made so many people leave their native land? Cabinet Minister Kock traveled through Hälsingland in July, 1845, to make a belated effort to reply to the question.[48] The answer is not simple. There was a complex pattern of motives, arranged in an order of priority which the questioner finds persuasive. The secular historian can find economic motives. Farmers were in trouble in Sweden during the years of the emigration. The crop looked good around Torstuna in the early summer of 1844, but then the rains came, day after day, flooding the fields and ruining the harvest. Spring seed sold in Västerås for the unprecedented price of three *riksdalers* a barrel. Potatoes were in short supply. On top of all this, the summer of 1845 was disastrously dry.[49]

Conditions were not better in Hälsingland. The flax crop, on which the prosperity of the province depended, was a failure in 1845. H. E. Ellsworth, the United States chargé d'affaires in Stockholm, wrote back to Washington on February 18, that there was "great distress in the northern provinces." There is no doubt that heavily indebted farmers were tempted to think seriously of emigrating to a country where rich farm lands were to be had for the asking. There were also concealed economic factors. Carl David Arfwedson, the United States consul in Stockholm, had played an important role in getting the Janssonists permission to travel. But he was also part owner of the Arfwedson and Tottie shipping line, the income of which was not diminished by a new flood of customers.[50]

It seems necessary to include economic factors in a full explanation of the reasons for the Janssonist migration, and no doubt a Marxist historian would find them decisive. The truth is that the Janssonists were as immune from economic blandishments as any human beings could be (except perhaps for monks). They were religiously committed in a sense which it is difficult for us to understand. The theory of the economic motive for their emigration is supported by the fact that, contrary to a commonly held view, few of the Janssonists were well off, and most of them were average Norrland Swedes and even less prosperous than Hälsingland folk in general.[51] All of them had to sell their possessions and put their money in a common treasury—a program with very little appeal for someone to whom economic motives were primary.

Equally plausible as a motive, and more difficult to document, is the theological utopianism which emerged inevitably from Janssonist presuppositions. If the genius of Lutheranism is its insight into the radical ambiguity of evil, making possible a truly prophetic confrontation of the contemporary scene, the genius of Janssonism was its sense of the messianic fulfillment. Probably the great majority of emigrants to America in the nineteenth century sentimentalized in some degree the utopian features of the new land—its natural riches, its social equality, its spiritual freedom, its promise of an economic cornucopia. But the Janssonists also had to find the great, good place where their perfectionism could flourish and where converts could be won. The Janssonists were utopians by definition, who shared with so many settlers of America the archetypal notion of the flight from the wicked Egypt, which was Europe, to the new Canaan, which flowed with milk and honey. The weakness, decay, and senility of Europe was contrasted in a way difficult for us now to realize, with the youthful innocence of America.

But these are general impulses, more or less applicable to all emigrants. The question which agitated state church leaders was whether or not the Janssonists were the inevitable offspring of the *läsare* movement, and if so, what steps could be taken to prevent its happening again. The difficulty in answering the question is that there were many kinds of *läsare*, and the word was used very loosely to describe any Swede who was serious about his religion. But the *läsare* could be recognized by a biblically oriented theology coupled with a Puritan ethical code. "When the reading of the Bible shows spiritual seriousness," wrote one student of the movement, "and one keeps the Sabbath day, and one for the sake of conscience abstains from swearing, card playing, dancing, drinking, then one can be called a *läsare*." [52] Was this type of piety directly responsible for the rise of Janssonism?

As has been mentioned, Baron L. M. Lagerheim, governor of Gävleborg province, was asked to find out, and sent three questions to every Lutheran pastor in his diocese. The questions were these: 1. How many of your parish are known to be of Eric Jansson's faith? 2. How many of these people belonged to the *läsare* movement? 3. To what degree may *läseriet* be said to have laid the foundation and prepared the way for this heresy, or on the other hand opposed it? The clergy reported 913 Janssonists in their parishes, 877 in Hälsingland and 36 in Gästrikland. The answers to the third question are too carefully qualified to make summary easy, but in general ten clergy thought the *läsare* were responsible for the Janssonists, and thirteen said that the *läsare* opposed Janssonism (at least after they found out more about what the sect taught). [53]

The truth seems to straddle both sides. Where the *läsare* were strong—in Forsa and Delsbo in the north; in Alfta, Bollnäs, Söderala, Ovanåker, and Voxna in South Hälsingland; and in Österunda and Torstuna, Uppland—it is significant that the Janssonists were also strong. But it is certainly true that when the solid type of *läsare* found out the fanatical and Separatist element in Janssonism, he not only became disenchanted with the movement but actively opposed it. J. E. Ekblom, who watched the origin and the growth of the Janssonists with discerning eyes, said this: "I think that a great many *läsare* were unaware of this [the true faith of Eric Jansson] until his catechims and hymn book came out, and these were not available here until the Janssonists left for America. If these absurd books had come into the hands of the emigrants beforehand, most of them would have seen through his heresy and stayed in Sweden." [54]

In the orgy of guilt and assessment of blame which followed the emigrations, the clergy were frequently chosen as scapegoats. *Aftonbladet* on February 8, 1845, blamed the pastors for the numerous Janssonists in Alfta. *Jönköpingsbladet* on July 18, 1846, said that the clergy had foolishly allowed the sects to become martyrs for the faith. *Söndagsbladet* on August 2, 1846, was unusually bitter against the ministry. According to the editor, the cause of the emigration was the shortcoming of the clergy, which, "itself without morality, and with an ugly life style, Jesuitical mentality, boundless indifference to duty, and growing disregard for solemn vows, let a sect blossom which is so crude that the inner stability of the land is threatened." *Hudkiswalls Veckoblad* on September 26, 1846, said the worldliness of the clergy left the door open through which the Janssonists burst. *Göteborgs Handels-och Sjöfarts Tidning* on May 7, 1849, blamed the whole sorry problem of the Janssonists on the weakness of the clergy.

Voices were not lacking to say that the inequities of the system of justice caused the sect to emigrate. It was especially noticed that when a Swedish citizen was found guilty of violating the Conventicle Edict, he lost his rights as a citizen, and could not testify in cases where he had been the victim of unjust attack. An anonymous correspondent to *Aftonbladet* on April 22, 1846, said that Sweden had lost some of its best farmers. "Recently," the writer said, "some fifty people, led by a bookkeeper, came to a cottage where four or five peasants sat and read the Bible. Some shook their rifles, some broke windows, and some cracked heads so that blood splattered on ceilings and walls." The writer asked why they did not take the case to the courts. "What good would that do?" the Janssonists replied. "We have tried for a long time to get justice, but we have always failed. If a hundred witnesses saw the violence they could not say so in court. They can't testify if they are *läsare.*" The writer said that after such treatment emigration was inevitable: "Persecuted by their distraught neighbors, refused due process and legal redress, denied even the elementary protection of life and limb, can anyone wonder why this desperate people seek a land where they can worship God in peace?" [55]

Posttidningen on November 20, 1846, printed without comment a letter from a Chicagoan sympathizing with the Janssonists, who had simply met "in peace and quiet to read the Bible, had their homes broken into, and were stoned, seriously wounded, and had their walls and ceilings stained with blood." *Dagligt Allehande* on November 27, protested. Printing such a letter, the editor said, without comment, suggested to the reader that the paper endorsed the criticism. On December 27, *Posttidningen* acknowledged the justice of the criticism, and admitted that it had serious reservations about the Chicago letter.[56]

The truth is that Sweden had become impossible for the Janssonists at precisely the moment when America began to emerge as a utopian dream in the European consciousness. There were enthusiastic travelers who brought back glowing reports of the new land. In January, 1842, *Aftonbladet* began regaling its readers with letters from Gustav Unonius, who loved this free and untroubled land. The letters were published in Norway: *Brev fra en svensk emigrant gavnlige öplysningar om de nordamerikanske förholde*, Christiana, 1843. Unonius was later to change his mind about the religious freedom in America, and was to return to Sweden; but in the 1840s he was a voice beckoning westward. Peter Cassel wrote glowing letters from Iowa which were published by *Östgöta Correspondenten*, and which told how free and equal Americans were.[57] He, too, learned to qualify his enthusiasm, and in 1849 wrote "this land can be at one time both a Canaan and a Siberia," [58] but his early views no doubt helped to establish the utopian myth. However, the immediate influence on the Janssonist emigration was surely Olof Olsson's fulsome letter.[59]

According to Eric Johnson, Jansson's son, his father was interested in settling in Illinois because of letters written home by Gustaf Flack.[60] Eric Johnson had excellent access to such information, but the book he wrote together with C. F. Peterson, *Svenskarne i Illinois*, is full of mistakes, and so we cannot lean too heavily upon it. The letters which influenced Jansson must have been written by Gustaf's brother, Carl Magnus Flack. We know that this Flack did indeed go to Gävle from Alfta in 1837 and had found work in a wholesale house. He and a friend, N. F. Åstrom, and four other young adventurers had set out for America in 1843, financed for the most part by the Gävle Trade Society. They settled in Chicago,[61] and seem to have sent home letters, now unfortunately lost, which are said to have spoken in complimentary terms of Illinois.

It is also likely that Lars Vilhelm Henschen had something to do with the decision to emigrate to America. Certainly he was an enthusiast about the new land and thought of it as a place where his ideal of religious freedom was embodied. He was a friend and disciple of the American Baptist preacher, Robert Baird, and could have caught from him some of the excitement of a free land which beckoned in the West. According to his sister, Augusta, Lars was something of a bore about America. "Henschen is crazy about America," she wrote in a letter to Maria Munck af Rosenschöld, "and wants to send everybody to that lovely land." [62] It seems in the highest degree likely that he advised Eric Jansson and Olof Olsson that their best hope lay in moving to the new land.

The temperance preacher, Robert Baird, must himself be considered instrumental in some degree toward the decision to emigrate. A graduate of Princeton Theological Seminary, he had traveled in Sweden in 1836, 1840, and 1845 and was appalled at the amount of drinking that went on. He had preached several times in Hälsingland, during his visit in 1840, and he was known as the author of a guidebook for emigrants written in 1832: *View of the Valley of the Mississippi, or the Emigrant's and Traveller's Guide to the West.* It may be because of this book that as early as November, 1845, the Janssonists were talking about settling in along the Mississippi River.[63]

Baird was also the author of *Religion in America; or, an Account of the Origin, Progress, Relation to the State, and Present Condition of the Evangelical Churches in the United States* (New York, 1844). It was a mine of useful information which the Janssonists desperately needed, and it helped the building of the dream: in America, said Baird, the government itself is Christian, a voluntary principle governs all religion, and the clergy do not need to have a formal education. The book was widely admired in Sweden. But the Swedish version, *Om Religionsfriheten i Förenta Staterna,* which is the only one the Janssonists could have used, was published in Jönköping in 1847, and so could not have influenced the Janssonists emigration of 1846. However, Baird was certainly a formative factor in the American myth, and his picture hung on the walls of many homes in Hälsingland in the early 1840s.[64]

Among the factors triggering the emigration to America must also be mentioned the Swedish newspapers, which at this time published letters and editorials praising the new land. Edward

Sjöberg charged in Stockholm's *Söndagsbladet* on July 5, 1846, that the paper *Aftonbladet* was a major cause of the emigration, since it regularly idealized America and whetted the appetite of peasants to travel there. Editor Palmblad of *Tiden* on September 4, 1847, repeated the charge. Lars Johan Hierta, editor of *Aftonbladet*, struck back on September 22, 1847, first admitting that there had been articles favorable to America, but adding, "to say that *Aftonbladet* through its stories of America has been the cause of the Janssonist emigration goes entirely too far." [65]

There can be no doubt that the periodical literature was in some important degree responsible for the folk emigration. To be sure, there were papers hostile to America, which published negative reports from settlers and played up the disappointments inevitable in the new land. This was the line of C. F. Ridderstad's *Östgöta Correspondenten*, *Jönköpingsbladet*, and Peter Fjellstedt's *Bibelvännen*. But other papers took a more positive attitude toward the United States, among them *Aftonbladet*, *Norrlands-Posten*, *Göteborgs Handels-och Sjöfarts tidning*.[66] A sea journey westward began to appeal to a great many people who were dissatisfied with their lot and believed in the possibility of historic fulfillment.

Some importance must also be given to the selections from the writings of the French traveler Alexis de Tocqueville which began to be published in the Swedish newspapers in the middle forties. His central theme was the triumphant joining of religion and freedom which he had found in America, and this report must have fascinated all Swedish sectarians who were writhing under the Conventicle Edict. Tocqueville was translated into Swedish and quoted at some length in *Jönköpingsbladet* (May 13, 1845), Uppsala's *Thorgny* (May 20, and May 23, 1845), and Gävle's *Norrlands-Posten* (February 4, August 1, and August 8, 1845). The Janssonists may have been especially intrigued by the February 4, 1845, issue of *Norrlands-Posten*, which published one of Tocqueville's chapters on "How Equality Inspires Americans with the Idea of Man's Perfectability."

Somewhere beyond the sunset, unimaginably distant geographically and in all ways a mystery, lay strange fields which were said to be fertile and largely uninhabited, where people could stay alive even if they could not speak the language, and where they could gather undisturbed to sing their songs, cry out to their God, and learn the meaning of Bible passages. They were not sad when they left, carrying with them bitter memories and also the bright hopes of pioneers, as yet undimmed by reality. An unknown bard in Sweden wrote a song, "Vi sålde våra hem," to describe what he took to be the melancholy of their departure.

> We sold our homes, and then we set out,
> As birds fly away when summer is done,
> The birds will return when springtime comes,
> But we'll not see our homes again.[67]

This may have been the mood of some Janssonists, especially of those who soon left the sect. But most of them had a more complex mood of anticipation mixed with apprehension, perhaps best expressed by the psalmist, "Rejoice with trembling." They had good reason for leaving. Time alone would tell if their new home would bring them the unqualified happiness for which they so much longed.

Land of Milk and Honey

> The concept mine and yours, with all that these words implied, would not exist for the Janssonists, but all possessions would be held in common. Was not this at least a lovely idea? But would it work in practice?
>
> —Victor Witting, *Minnen från mitt lif som sjöman* (1904)

Some twelve hundred Janssonists sold their possessions, packed what they thought they could carry in homemade chests, and trundled down the roads to the port cities of Gävle, Stockholm, and Gothenburg during the next few years.[1] They boarded vessels which were built to haul cargoes of lumber and iron, and on which the passenger accommodations were minimal, and they found themselves crowded together in poorly ventilated and unhygienic holds. The perils they faced were plain enough: probably permanent separation from families left behind and from their ancestral setting, seasickness, scurvy, storms and shipwreck, theft, death by drowning, or disease. The danger was real enough to make a strong man blanch. But these emigrants had without planning it that way equipped themselves with the best possible protection against suffering and danger: an unshakable faith in their errand. They believed without a moment's doubt that suffering was not the prelude to disaster, but on the contrary the prelude to glory. Hard times were the signs that they were true disciples of their crucified Lord. They were used to trouble, and they were ready for more as they set out to give their dream a local habitation and a name.

The plan was to build a holy city like the one John saw descending from the clouds like a still unravished bride. The details were not clear, but they knew that with their common funds they could buy prairie land which could be had for a song; and on this land they would improvise houses and also have a place where

Capt. Eric Johnson (1838–1919), son of Eric Jansson, was a newspaper and magazine publisher. Painting by Olof Krans, courtesy of the Bishop Hill Heritage Association

they could worship God in the morning and after their day's work in the rich fields. No sheriff would harry them, and their meetings would be publicly announced. Their beloved prophet would explain God's word to them, and they would sing the pilgrim songs with grateful tears in their eyes. The world would see springing into being on the grasslands of Illinois an indisputable miracle, a stage for the drama of sinlessness, the enfleshment of the idea of separated innocence.

The difficulty which some of them may have guessed when they leaned on the ship's rail and watched the gray Atlantic slipping by is that heavy-footed reality cannot really reproduce the pirouettes of a dream. Building a holy city is a richly contrived process, involving economic decisions, psychological and sociological complications far beyond their simple certainties. Yet no one needed to tell these men and women with the marks of work on their hands that their vision of sinlessness had somehow to take form in the actual world if it were not to disintegrate. The trouble was that the whole emigration—the closing of affairs in Sweden, the setting out from the harbor, the sea voyage itself—began to seem to some of them less and less like a pilgrimage of pure people and more and more like the struggle of troubled people through Luther's world with its strange mixture of good and evil. They may have begun to see that no single human act—certainly not the transplanting of twelve hundred assorted people in a strange land—could express in its simplicity the vision of purity which had captivated them in the farmhouses of Hälsingland.

Still they had Olof Olsson's hearty optimism in their minds. He had arrived just before Christmas, 1845,[2] and it seemed to him symbolic that an American child of about eight years had taken him by the hand and led him to a haven, the Bethel ship *John Wesley*.[3] That spring Olof Gustaf Hedström had begun his ministry in the great port of entry, and he had already established his reputation as a wonderful host, warm, unhurried, courteous, and devout. No one traveled the Erie Canal to Buffalo during the winter time, so the Olsson family spent the winter on the Bethel ship, and came deeply under Hedström's influence. At one of the services on the Bethel ship, Olof Olsson read the hostile permission to travel given him by Pastor Scherdin in Söderala, and the congregation burst out in loud "Hallelujah!" Olof wrote a letter telling all about those cries to his brother Jonas, in care of Pastor

Mathilda Jansson, Eric Jansson's daughter (1842–1926). The family tradition is that she looked more like her father than her brother, Eric. Portrait by an unknown photographer, courtesy of the Bishop Hill Heritage Association

Scherdin. To Jonas's embarrassment, Scherdin asked Jonas and his wife to sit on either side of him while he read the letter aloud, including the account of Scherdin's caustic remarks in the permission to travel.[4]

Olof Olsson enjoyed his winter in New York, and in fact decided to join the Methodists. The theology preached by his genial host was so much like that preached by Eric Jansson that only a theologian with nothing better to do could tell the difference. But of course he had an errand to carry out, and when the canal boats were running again he and his family set out for the interior. Though he had at first planned to look over Minnesota and Wisconsin before deciding on the best setting for the Janssonist colony, he was in fact disposed to settle in Illinois. The lands bordering the Mississippi had already become enshrined in mythic setting by the writings of Robert Baird. Olof Hedström recommended the neighborhood of Victoria, Illinois. His brother, Jonas J. Hedström, had fallen in love with an American girl in New York, Diantha Sornberger, and when her family moved to Victoria, Jonas followed along. He married Diantha, settled down in a log cabin while carrying on a career as Methodist preacher and blacksmith, and wrote back enthusiastic letters to his brother Olof in New York. Jonas was a happy pioneer. Having lived in America since 1833, he had forgotten how to read Swedish and had to ask for help from a newcomer when a letter came from Sweden.

It was in Jonas's log cabin where Olof first stayed when he arrived in Victoria. Jonas had a theology much like his brother's, and much to Olof's liking.[5] In other respects he was unlike his fastidious brother, taking on the manner and appearance of the frontiersman. He wore faded clothes and stuffed his pants down into his boots. Himself a farmer, Olof Olsson got along very well with Jonas, and the two traveled about looking for the best spot for the Janssonists to buy. They decided on a fertile spot on the crest of a low hill a few miles north of Victoria, in Henry County, where there was running water and also an attractive grove of trees, surrounded by endless fertile prairies. His eyes popping at the deep rich loam, Olof also bought a forty-acre tract for himself and seemed to have decided to strike out on his own.

Eric Jansson arrived in New York early in June together with Charlotta, and with his two children, Eric and Mathilda. He spent several weeks in New York, unlike Olof Olsson arguing a great deal with Olof Hedstròm, and holding meetings of his own in the Swedish houses nearby. According to C. G. Blombergsson, he even won several converts to Janssonism in New York, including some who could not understand a word of Swedish but who responded to his obvious conviction and persuasive tone.[6] Some Swedes were also won over, including one Sophia Pollock, who was to play an important role in the later life of the colony. She had an unusual career. Born a foundling in Gothenburg, and growing up as a very attractive young lady, she had married a sailor at an early age and shortly afterward was left a widow.[7] Coming to New York, she married a Mr. Pollock and persuaded him to attend some of the Janssonist meetings. She became convinced of the truth of their teaching, joined the sect, and took an especial liking to the little Eric, then seven years old. Eric liked her, as well, and even helped her teach some of her classes. The Pollocks accompanied the Janssons when they started out for Victoria. On the long journey westward Sophia served as translator and tutored the family in English.[8]

Indulging his penchant for predicting the future, Eric Jansson said that all the Janssonists would arrive in America by New Year's Day, 1847. But the Atlantic, unimpressed, did not cooperate, and many ships were delayed in route. The brig *Sophia*, out of Gävle, arrived on January 7, and the bark *Augusta* did not get to New York before March 8. On March 20, the bark *New York* arrived with a large part of Janssonists aboard. Anders Andersson and John Björk came from Bishop Hill to welcome this party and to escort them to Bishop Hill, and with them as interpreter was a former sailor named Hans Hammarbeck.[9]

The threat felt by many colonists was that they would be cheated by the Americans, overcharged for passage on canal boats, overcharged for food, and swindled out of their possessions. All these things happened, but they were sometimes victims of their fellow Swedes as well. A. M. Ljungberg, the Bible salesman from Uppsala whom Anna Maria and Maja Stina had met when they went to see the king, had become one of the early Janssonists to leave Sweden and had brought his family over using common funds. As treasurer of the group, he kept the communal funds in a locked chest. When the time came for him to pay for the passage westward to Illinois, he refused to part with the money. Fortunately, Anders Berglund, Anders Andersson, and Hans Hammarbeck were in New York at the time. Biding their time until Ljungberg was parted from his chest, they pried it open with a nail and took possession of the communal funds. Ljungberg was furious when he discovered what they had done.[10]

There were other shocks which threatened to destroy their utopian illusions. One was the frequency of illness among them. Eric Jansson had promised them that if someone fully trusted in God, he would never be ill; yet these farmers suffered mightily from seasickness before they found their sea legs. Scurvy was not uncommon, since their diets were inadequate during the long sea voyage, and their flesh sometimes seemed to be rotting away.[11] The baker from Bollnäs, "Bröd-Jonas" Malmgren, went about amongst the sick as a chaplain. But his ministry did not help them, because he scolded them for having so little faith that the devil could take possession of their bodies.

Their other major source of disillusionment was their inability to speak the English language, as Jansson had promised them, the moment they set foot on American soil. Olof Olsson was the first to experience this fact when he stood speechless on the dock at New York, even though L. V. Henschen had given him an English-Swedish lexicon to use when talking to Americans. In his letter back to the Janssonists he said that his biggest problem was that "I have not been able to ask for work and be understood." [12] Brita Gustafsdotter, who had counted heavily on this miracle, found her inability to speak English confirmed her growing belief that Jansson was a charlatan, and she left the colony. So did Eric Jansson of Myrsjö.

Many people defected from the Janssonists in New York and in Chicago, and the story gained currency that many had simply pretended to be Janssonists in order to get free passage across the Atlantic.[13] In any case, some of the colonists were penniless and had to be supported by the others; some were soldiers who were released from military service by the payment of communal funds; some families were in debt and could come to America only if the others paid the debts. And when some of these people broke away without so much as a thanks to the colonists, the Janssonists were understandably bitter. Anders Larsson and his family led the deserters who stayed in Chicago, some twenty-seven strong.[14] He wrote to Sweden and said that among the disaffected were Jan Jansson from Vannsjö, Pehr Ersson from Grinda, Petter in Mellanbo, and a number of folk from Hälsingland.[15] Anders Andersson explained that he had had enough of perfectionist theology, and after listing its faults, summarized his complaint: "God save us from Hälsinglanders!" Probably the defection which shook Eric Jansson the most was when his brother, Jan, left the colonists and settled in Chicago. Many of the defectors lived at first in a house on Illinois Street between Dearborn and State streets.[16]

There were disgruntled people who left, but the general mood of the Janssonists was good, despite their harrowing experiences. A Chicago Swede wrote to the editor of *Najaden* (Karlskrona) that on September 3, 1846, he had seen the party from Dalarna, sixty-five strong, camping outside his window for three days. Though it was ninety-eight degrees in the shade, he found them patient, understanding, and cheerful. The correspondent had talked with their guide, one Tolk, a gentleman who had been an officer in the Swedish army. Tolk related, "I joined the Janssonists not because I believe in Eric Jansson, though he is a remarkable man, and his knowledge of coming events is astonishing; but I saw these poor people, good, sincere, and upright in their religious faith, driven from their fatherland by persecution, compelled to travel among strangers without knowing their language. I hate persecution, and so I decided to take care of them." [17]

In the late summer of 1848, Victor Witting, the sailor who had been won over to Janssonism and had followed them until he had fallen ill in Chicago, said that he was delighted to see a party of Janssonists arrive on their way to Bishop Hill. Anders Andersson, one of Jansson's chief assistants and one of the abler apologists, arrived to serve as a guide to lead the newcomers to Bishop Hill. It happened that Unonius was also in town at the time, and had announced a meeting for that evening—no doubt the first formal Swedish service in Chicago—at the Medical Institute on the north side. Witting and his Janssonist friends decided to attend. Unonius was wearing his Anglican cassock, and preached to the group in his restrained and cultivated manner. After the sermon, he invited the congregation to ask questions, and Anders Andersson began to argue with him about the biblical basis for his Anglicanism. According to Witting, Andersson was clearly the winner.[18] Unonius might well have lost this kind of argument, and the fact that the exchange is not mentioned in his memoirs may be proof that he preferred not to remember it. He did put down his recollection of a similar encounter later on in Galesburg. After hearing him preach, he said, the Janssonists surrounded him: "Anyone who had ever come in contact with that sect knows how fruitless it is to converse with them or to convince them of their error. Furthermore, when I discovered their tendency to become irritated, I sought to get away from them; but they pursued me with their shouting clear into the street." [19]

Eric Jansson and his party—composed of Maria Christina, Eric and Mathilda, a sister of Sven and Louis Larsson of Söderala, an old lady from Falun, and the Pollock couple—had arrived in Victoria in July 1846. They had followed the usual path westward from New York: steamer to Albany, canal boat to Buffalo, propeller-driven boat through the chain of the Great Lakes to Chicago, canal boat to Peru, and the final lap by horse and wagon. Many years later Eric's son, Capt. Eric Johnson, wrote an article for his own magazine, *Viking*, in which he reminisced about the first days in Victoria. He said he could not remember where the Falun ladies stayed, but the Pollocks were put up in a nearby cabin. The Jansson family stayed with the Olof Olsson family in Jonas Hedström's log cabin, which was one day to achieve a modest fame as the birthplace of Swedish Methodism in America. The captain recalled that snakes came through chinks in the wall and sometimes were found in the bunks. One night the rain came down so heavily through the roof that they all huddled under an old umbrella.[20]

But what the captain recalled most clearly were the violent theological debates of those first few days. Jonas Hedström and Eric Jansson had much in common: both were Swedes, passionately religious, pioneers in a strange land, and advocates of theological systems that sounded to an untrained ear to be exactly alike. L. P. Esbjörn, the Delsbo youth who was one of the early Lutheran pastors in Illinois, said this of Jonas Hedström: "He often preaches the same as Jansson, that the root of sin must be pulled up in this life, and that one must be free of sin if one would be holy. He denies that sins of weakness can be found in the faithful. He is wholly unlike Scott (who was a *Wesleyan* Methodist), and I cannot find much in him other than the carnal desire to get members on his rolls."[21] There was a plan that Hedström and Jansson were to hold a formal debate over a period of three days in 1847, but after one day of bitter name-calling, the whole exchange was called off.

There was beginning to be an intense rivalry between various confessional bodies, each of whom wanted the support of the newly arrived immigrants, and it was this factor, rather than any other theological difference, which strained the relationship between Jansson and Hedström. But there were other factors: Hedström had the view that each pioneer should own his own home and till his own field, while Jansson was promoting a communalism which made the interest of the group paramount. Olof Olsson, whose conversion to Methodism had already begun under the tutelage of Olof Hedström in New York, now found himself further convinced after meeting the rough and hearty blacksmith pastor, Jonas Hedström. He had already begun cultivating his own land when the Janssons arrived in July.

Captain Johnson remembered his father's anger when he found out that Olof Olsson, his advance agent, had strayed from the true faith. The others listened silently while Eric and Olof argued theology during breakfast, dinner, and supper, as well as between meals. Eric hid under the blankets one night when he heard the massive Olof Olsson threaten to throw the smaller Eric Jansson bodily out of the cabin, but he was pleased to find out on the following morning that the two men were still friends.[22] It seems likely that Olof returned to the Jassonist faith. On August 1, the colony purchase of an eighty-acre farm at Red Oak Grove was made over his signature. But he stayed on his own land, and before the year was out he and his wife and their two children had died of cholera.

Before the end of 1846, the colony had grown with the arrival of each ship, and soon there were four hundred living at the settlement, of whom seventy lived at Red Oak Grove, and the others at Hopal Grove.[23] Picking a name which gave away their nostalgia, they called the new village Bishop's Hill, after Eric Jansson's birthplace, Biskopskulla. In time the *s* dropped away, and the village was simply Bishop Hill. There was a strong feeling of unity between the central village and the outlying farms. There are still traces to be found of the old ox road, which ran from Red Oak Grove through Buck Grove to Bishop Hill, and on which there was a steady traffic of colonists on various errands or bringing loads of oak and maple from the saw mill.[24]

The first homes were the log cabins already existing on the land, such as the one in which Linjo Gabriel Larsson lived when he arrived ill from his journey. When the large parties came in the fall, other emergency measures had to be taken, and so twelve dugouts were built on both sides of the ravine, having log sides and earthen backs, with two windows in the front. In one of these, "Klumpstugan," lived fifty-two young ladies. A red frame house was built for Eric Jansson and his family during the second summer, but later he moved to another house which had better bathroom facilities. At the end of his life he had rooms in what was called "Big Brick." [25] The church at first was a structure built of logs in the form of a cross, and covered with canvas and skins, in which eight hundred people could worship seated on logs. When parties of immigrants arrived, they frequently slept there until more permanent quarters could be assigned. But the chief function of the tent church was to provide a place for the daily morning and evening services and the three services on Sunday. Early settlers tell of Jansson himself walking down the row of dugouts at five o'clock in the morning, rousting people out of bed to attend the early service. But at Christmas, 1846, a brass bell was secured, and all could rejoice at this proud summons to public service. No longer need they meet secretly in the Hälsingland farmhouses! Unfortunately, a careless colonist let some sparks fall from his pipe into a pile of dried flax, and the fire that followed burned down the church as well as a nearby tailor and weaving shop. Plans were immediately set in motion for the present Colony Church, and this was completed in 1849, its walls made of adobe brick strengthened with straw.

Very quickly there were erected three adobe kitchens and eating halls, the east one for the Forsa, Västmanland, and Söderala colonists, the middle one for those from Alfta and Ovanåker, and the west one for the people from Dalarna. In the summer of 1848 a neighbor, Philip Mauk, a Hoosier who settled nearby, taught them how to make kiln-dried brick using local clay,[26] and in a flurry of construction they built a bakery, a brewery, and a flour mill on Edwards Creek. They even had the audacity to build by far the largest building on the central plain at that time, Big Brick, forty-five feet wide and two hundred feet long, with a basement and three floors. Very like the "grand phalanstery" of the Fourier utopian colonies, Big Brick provided ninety-six rooms besides a large hall for dining.

Farming was of course the chief occupation of the colonists, coming as they did from rural Sweden. They grew great quantities of Indian corn, harvesting that first crop with scythes and beating the grain against barrels. In 1848 they used cradles for the first time, and in 1849 they had their own reaper. In time, broomcorn came to be a chief source of income. To their delight, they discovered that flax would grow in Illinois as well as in Hälsingland, and this crop came to be a mainstay.[27] They raised hogs, cattle, and sheep, chickens, turkeys, and geese. The younger boys had charge of the oxen, the older boys took care of the horses, and the girls cared for the cows, hogs, and fowl. The marks of civilized life began to appear in this primitive setting, as they did when Robinson Crusoe put together his island home. The whole enterprise had the busy and happy quality of a medieval village, and Helen Lindewall remembered in later years the sound of several hundred laborers marching home from the fields in the evening, two by two, singing Swedish folk songs.

There was some industry in the colony, as the resourceful settlers cast about for some form of trade to supplement their farming income. All the utopian colonies devised things to sell, and the Shakers, who had some communication with Bishop Hill, were especially ingenious, producing gadgets to core apples and machines for washing sheets. The Oneida perfectionists grew, canned, and sold farm produce; ran a busy sawmill and flour mill for their neighbors; and sold chairs, silk thread, and community plate which is still on the market. The ladies of Bishop Hill wove endless yards of linen cloth, twelve thousand yards being counted by 1847, and the quality of it was comparable to the famous woolens of the Amana colony. A hotel was built to provide rooms for stagecoach travelers coming from or going to St. Louis.

The result of this devoted communal effort was a growing productivity, and since the overhead expenses were minimal, wealth began to accumulate as it did for the other utopian groups, such as the Quakers. Four men were in charge of the communal treasury: Jonas Olsson from Ina, Olof Jansson from Valla, Olof Jonsson from Stenbo, and Anders Berglund from Alfta. The character of the community began slowly to change from that of poor and pious settlers to reasonably well off burghers. When Daniel Lonberg visited the colony in 1849, he noticed that some of the old piety had withered away, and that a new capitalistic hardness had begun to take its place. "All this bears proof," he said in a letter home to Sweden, "of the fact that they now pay less attention to religion and more to industry." [28]

Some of the inner attitudes were of course the natural response to changing circumstances. Eric Jansson took the leadership in shaping a new life-style for the community which expressed basic theological positions and yet made some recognition of the special demands of time and place. The basic pattern of conduct was what they took to be that of the primitive church immediately after Pentecost. Some of the life-style was invented, some was born by the stark necessity of survival, and some was imitated from other colonies, such as the Shakers.

Reacting against what they took to be the unwarranted intrusion of state officials in their devotional life, the first principle of the Bishop Hill Colony was security from outside interference. The impulse was monastic in its inspiration, however much that would have offended Eric Jansson had he been told it. The logic of the society of pure believers uncontaminated by the world pointed to just such a separated place as Bishop Hill. But it is a utopian ideal as well, all such colonies seeking to protect their identities by minimizing distractive elements from outside. Of course the immigrants were glad to band together in any case, in order to preserve ancestral ties in a land of strangers, and to give mutual support against predatory foreigners. Notions of unity and separation have been shown to be immigrant traits, and these motives were strongest of all amongst sectarian immigrants.[29]

The principle of seclusion was to prove difficult to maintain in the midst of a frontier land bursting with opportunities for young people without their parents' theological scruples. It was also certain to breed suspicion on the part of the neighbors who were not very clear about what the foreigners were up to, and this factor was to cause the Bishop Hill folk a great deal of trouble. In any case, the colony practice was clear. There was from the beginning a principle of voluntary entrance, anyone being invited to join who could reply "yes" to the query, "Do you want to be holy?" There was also of course the right of voluntary withdrawals, and a great many people used this privilege. But when the ties with the colony were broken, two penalties came into play: one was that he who left must suffer the Prophet's denunciation, and make his exit like Judas sneaking off to hang himself. The other was even more difficult to bear: when a person left the colony, he also left all the money and labor he had invested in it and kept for himself only the clothes he had on his back.[30]

Much of the criticism to which Bishop Hill was subjected by disgruntled colonists who wrote back to Sweden must be attributed to this policy. The point may have been agreed to by neophytes, who when they joined the colony had no interest in the conditions of departure; but when they left and found themselves insulted and penniless, tempers flared. It seemed not to be consoling that this principle was not unique at Bishop Hill, and was no proof of the special cruelty of the colony leaders. Totalitarianism was the common practice among the utopian colonies. For example, the Articles of Association of the Harmony Society at Economy, Pennsylvania, which had been in force since 1805, stipulated that money and material goods would be jointly owned, and that if any member left, he forfeited his claim to these possessions.[31] The point was that the stability of the colony could not be threatened by members withdrawing their contributions to the common fund whenever they felt like it; and the absence of any such right was also of course a strong inducement to stay, despite disappointment or even disillusionment. Jansson thought of his colony as being a unity managed with authoritarian power; and the analogy to a medieval village was made explicit by the erection of an earthen wall which circled the colony for four and one-half miles, holding in those who were inside and holding out the nonbelievers. Traces of this wall may be seen today.

Related to the idea of seclusion, defining the separation from the world, was the idea of communalism, or mutual property within the sect.[32] The Bishop Hill Colony members were to be mutually dependent, helping each other like a circle of covered wagons defending against an Indian attack. There were to be common meals, though the men ate separately from the women, and the children from both. They were to dress alike, and clothes were drawn from a common storeroom. Work was to be shared, each one doing what he could according to his training or talent, and all pitching in for the seasonal work in the fields. They lived together in what might be the first apartment building in the Midwest.

The communalism which they practiced was primarily prudential, rather than a doctrinal rejection of private property. Anna Maria Stråle explained the basic idea to Johan Sagström, in a letter written on March 5, 1872: "For each one of us individually to construct a home with the few assets which we brought over, after paying for the trip, was impossible. So we decided to run farms in common." [33] To be sure, the practice was Pentecostal. The early church, as described in Acts 2:44, was at least in some places communistic: "All that believed were together, and had all things in common—and were of one heart and one soul." Finally, there were examples of the communalistic system proving practical by such utopian colonies as the Shakers, the Amana perfectionists, and the Hutterites. The Bishop Hill communalism was strictly temporary and was in fact abandoned in 1853 under the pressures of a growing individualism.

Another feature of Bishop Hill as it developed under Jansson's leadership was the claim of power over physical disease, a boast also of the Shaker colonies. The promise was first made in connection with seasickness, and the first doubts of its truth sprang up at sea. On land the colonists were decimated by a series of fatal illnesses, brought on by fatigue, overcrowding, poor hygienic conditions, inadequate diet, and scanty medical care. There was illness, but the theory could remain unshaken, since the leaders could always say that if anyone felt sick, he simply gave visible proof of his lack of faith. Many letters by deserters pointed to the callous treatment which was given to sick and dying colonists. But what looked like plain cruelty was really an unwarranted confidence in faulty theological analysis. It was supposed that when a believer really took Christ into himself, his body could no longer serve as a host for physical disease.

The Janssonist position on marriage and sex had as little of the doctrinaire about it as communalism and was itself a response to the exigencies of frontier life. In Sweden a married couple who joined the sect stayed married, and nuclear families were kept intact. But no new marriages were performed. In the new land there was enough to do to house, feed, and clothe adults, without bringing any more helpless babies into the world. For the first year they tried out celibacy at Bishop Hill, though their instinctive sympathies were rather with such utopians as the Icarians in Iowa, the Amana colony at Buffalo, the Aurora colony in Oregon, the Bethel colony in Missouri, and the Zoars of Ohio—all of whom honored the family more than the single life.

The celibacy of the first two years was surely not a formula which promised a bright future for Bishop Hill, especially when the flow of new members dried up. The Shakers, who were descendents from the English Quakers, thanks to their verbal facility were able to make converts all over the country, and in their case, celibacy was not a recipe for extinction in one generation. Neither had celibacy this effect in the Catholic tradition, since celibacy there is not proposed as a universal virtue, but is one of the "counsels of perfection," intended for ordained clergy.

The Janssonists had no such carefully worked out theory of celibacy, but only a vague hostility toward sex, expressed in the separate living quarters, separate meals, and separate sections for men and women in the Colony Church. A policy is no less irritating because it is vacillating and indecisive. Young people especially chafed under the prohibition of marriage, and not a few suffered from *giftaslysten* ("longing for a wedding"). As a result, Jansson had an abrupt change of heart and announced to his followers in June, 1848, that marriage was not only permissible but highly pleasing in the eyes of God, who after all had ordered men to people the earth. Big Brick was ready with its many new rooms.

The consequence was a sudden rash of weddings. "Immediately," according to Daniel Lonberg accustomed to the slower Swedish custom of publishing bans, "there was a shocking pairing off between men and women, like and unlike. Any woman a man wanted he could take. Women had no voice in the matter." [34] The marriage records in the Henry County Court house at Cambridge, Illinois, reveal that Jansson married four couples on June 25, 1848, three couples on July 2, four couples on July 9, five couples on July 16, twenty-four couples on July 23, and nineteen couples on July 30. There was no elaborate ritual. The bride simply carried a bridal wreath, and after the ceremony the couples marched in a procession down to the mill. There followed a bridal feast of turkey in the dining hall of Big Brick.

Apparently not all of these marriages, as Lonborg said, were the result of romantic ardor, but were soberly plotted to suit the interest of the colony. One day Catharina Magnusson (called "Tovås Mor" because she was from Tovåsen, Färila Parish) was busily plying her trade as a bricklayer and was building the Colony Church. A messenger came to her saying that Eric Jansson wanted to see her. Without washing her hands, or even removing her mason's apron, she left immediately to see what her leader wanted. His idea was that she should marry at once, and he had a bridegroom in mind: Paul Myrtengren, who had a harness shop in one of the dugouts, and who languished as a widower at forty-eight.[35] She agreed, but whether with joy or dismay we shall never know. When she returned to her job, she found that Eric Linbeck had laid bricks for her during her absence. She told him what Jansson had proposed, and thanked him for his help, whereupon he replied that this was his wedding present for her.[36] The Henry County marriage records note the fact that Paul Myrtengren and Catharina Magnusson were married on July 23, 1848.

Despite the "milk and honey" phrases that were sent back to Sweden, the colonists did not fare sumptuously. Fasting was part of the life-style. Helena Lindewall remembers that during the fall of 1847 they fasted every Sunday, consoling themselves only with a little "dricka," the local small-beer. There were no ascetic implications. Fasting as a penitential discipline implies a strong doctrine of sin, and since the Janssonists could confess to no sin, they had no sense of penitence, and so no need of its symbolic expression. Fasting at Bishop Hill relieved the larder, rather than anyone's conscience. The project of feeding eleven hundred people three meals a day under pioneer conditions would appall the most stouthearted cook, and it is not surprising that on some days fasting was the only option. The colonists ate for most meals cornmeal mush, "tunnbrod" and "pölsa" (Hälsingland specialities), and on rare occasions meat and poultry. Very seldom did they have real coffee, and the flour was sometimes made of slippery elm bark.

As has been pointed out, the Janssonists never managed any great enthusiasm for teetotaling, and their Lutheran and Methodist neighbors, more puritan in their taste, were often shocked at what they regarded as abandoned tippling. In this respect they were not unlike some of the other mid-nineteenth-century utopian societies with European backgrounds: the Harmony Society

Rappists were famed for the quality of their whiskey; Amana, Economy, and other German communities drank wine and beer rather than water; and the Economy Society had large wine cellars. One of the Janssonist's first buildings was a brewery, and they drank at most meals a nonfermented drink called "svagdricka" (near beer); but on special occasions and for medical purposes they also had a stronger concoction called "Number 6," which had an alcoholic base and was laced with cayenne pepper.[37] Like the Aurora and Bethal colonies (but unlike the Icarians, Harmonists, and Zoar Separatists), the Janssonists had no quarrel with tobacco. Jonas Hedström and his Methodist congregations had managed to develop a perfectionist theology which at the same time had room for carnal sins, but the Janssonists never achieved this.

Part of the Janssonist life-style was a strong purpose of missionary expansion, but the American soil seemed very stony for this kind of sowing. In the preface of his *Catechism*, Jansson expressed his intention to "go forth into all the world, though all the world's kings and princes should fight against the Gospel, day and night." Traveling in disguise and hiding under barn floors, he nevertheless thought of his mission as encompassing the world. He would perhaps have been only mildly interested to hear that his program was very like that of Francis Bacon in the *New Atlantis*, in which King Solomon forbids strangers to enter, or natives to leave, except that twelve "merchants of light" were to sally forth and convert the world.

In 1847, Jansson set up a school for training these apostles, and it met like a band of Druids under an oak tree and in a dugout at Hopal Grove. The idea was that when they were thoroughly grounded in the biblical basis for perfectionism, they would fan out like Mormon zealots into the wide world—into neighboring states, and into Sweden, Norway, and France. Probably because of administrative burdens of running a large colony and at the same time serving as the whole faculty of a theological seminary, Jansson was in the habit of asking visiting clergymen to address the school. One such visitor, "S. B." as he signed himself, seized the opportunity to chide the apostles for leaning too heavily on Jansson, and expounded the text: "Woe unto those who rely upon man." He heard afterward that Jansson was so annoyed by his message that he spent the next few weeks commenting bitterly on it.[38] When the flow of information dwindled from the faculty, and when the flow of the Edwards Creek slowed to a trickle, the apostles were drafted to walk inside the huge waterwheel which kept the great millstones turning.

The project of winning the world to Janssonism died aborning. There was an initial confusion of policy, not unlike that which befogged the Fourier experiments, which tried to achieve a monastic exclusiveness and separation and, at the same time, launch a furious effort to transform and penetrate the world. Catholicism had solved that riddle, if Jansson had only noticed, by assigning monastic withdrawal and missionary expansion to different agents. Language was also a terrible problem for the Janssonist missionaries, who could be persuasive only in Swedish. Daniel Lonborg claimed that they had not a single convert.[39]

There was, however, some communication between Bishop Hill and other utopian colonies. There were three other utopian communities in Henry County at the time: colonies at Morristown, Wethersfield, and Geneseo were thriving, though one had failed at nearby Andover in 1835. There was a school for missionary clergy at Galesburg, run by George Washington Gale, which became the nucleus for the antislavery sentiment in the Fourth District. But Bishop Hill seemed unrelated to these settlements. The tailor Nils Hedin did, however, visit the perfectionists at Oneida and the Rappists in Pennsylvania. The Shakers sold the colonists some prize Durham cattle, taught them how to make brooms of cornstalks, how to dye wool and weave cloth, and how to grow fruits and vegetables. And when Nordhoff visited the Pleasant Hill Colony in the 1870s, he found "a good many" Swedes living there, most of whom came from Sweden, but some of whom might well have been former Janssonists.[40]

Anders Blomberg, the eccentric tailor from Färnäs, Dalarna, and Olof Stenberg, from Stenbo, had been sent out by Jansson on a missionary journey in the spring of 1848, and reported back to him from Pleasant Hill, Kentucky. They had started south to Hillsboro, and had received a cold reception. Their brave effort to write English adds to the letter's charm: "It was so vary bade report above our colony, and they spoke very hard against us in the beginning, but still before we leave them we been invited to preach." They had passed through Vandalia, Salem, Fairfield, and Mount Carmel, speaking in various places. Louisville seemed to them like a new country, but the six Swedes they found there gave them a rough time, "especially some when they heard that it was we that hath burned the books in Sweden." But "when we gave our explanation they most thing over it, that we was not crazy." [41] One of the Swedes they met who had the unlikely name of George H. Collini, said that he had once worked for E. Trolin in a store at Ovanåker, and that he had considered moving to Bishop Hill. Actually he later moved to Louisville.[42]

Anders and Olof moved on to the Shaker colony at Pleasant Hill, Kentucky, and were pleased to find the same perfectionism that Eric Jansson preached being taught there: "They shall live perfect lives holy as Christ Jesus for he that is born of God can not sin and all other denominations in the world they call anti-christ. . . . And us they receive like brethen. But what should we say when we meet such a trumpet or voice?" Their hosts failed to convince the missionaries that there was biblical backing for Shaker views of the Second Coming; [43] but there was an agreement to continue the discussion: "They have said . . . that they shall go to our colony, and tri [*sic*] if we have speak the truth. They believe us but they will like wise go there. Therefore may ye be prepared to entertain them." [44]

When the two missionaries returned from their travels, they gave the colonists a mixed report on the virtues of celibacy. Anders Blomberg had become convinced that the Shakers were right, and that a truly spiritual life did not have room for sexual relations. He preached in the Colony Church on this subject, and the matter was much discussed. But the colonists were not markedly ascetic in their points of view, and were not ready for such grim sacrifices and dubious rewards. On the following night, Olle from Stenbo gave his report in the Colony Church, and this sermon turned out to be a song of praise for the Christian family. While he preached, Anders Blomberg sat silently in the gallery with his legs draped over the bench in front of him. The next morning he packed his bags and left for the Shaker colony at Pleasant Hill, Kentucky, never to return.[45]

In addition to the school for apostles, the Bishop Hill Colony also started during the first year a school for illiterates, since more than half of the colony could neither read nor write. At first it met in the church tent and was taught by Margaret Hebbe, while another school at Red Oak was taught by Karin Petterson and

Mrs. Christina Ronnquist. By the first Christmas, Mrs. Pollock was teaching a class in English in one of the dugouts. In January, 1847, a Presbyterian minister, the Reverend Mr. Talbot, taught the colony children English, and in July he was succeeded by a medical doctor, Nelson Simons.[46]

Despite the expected and unexpected reversals, the story of the Janssonists for the years 1846 and 1847 must be considered a happy one. The letters written home by colonists who stayed sound the same note of thanksgiving: the trials had been many, but they were as nothing compared with the new joy that had been found. A letter from Anders Andersson to Bailiff Ekblom is typical. In America, he writes, "every one has the right to serve God according to his conscience. Here we can listen to Erik Jansson expound the Bible without being disturbed by anyone. The light that God has lit in Erik Janssen cannot be hidden either in jail or in the lunatic asylum." [47] The editor of *Dagligt Allehande* complained sourly that the Janssonist letters sounded so much alike that they must have been written by the same hand: "gold grows like potatoes on the farms in America" was one of these repeated expressions.[48]

On February 10, 1847, Eric Jansson wrote a gloating letter to Alderman Henschen: eight months ago, he said, he had left behind "all European powers and dominion. In this free land, no one comes to wrench the Bible out of my hand as the Administrator of Forsa County did on Midsummer Day, 1845." He did report an ominous note which curiously foretold his own death: "When the time is ripe . . . our blood will flow for the sake of the truth in this land of freedom." He had been studying the English Bible, and was pleased to find his own perfectionism even more clearly supported in the King James translation: "How Satan had blinded the Lutheran clergy who claim that the Bible in its original language or other language makes it appear that the roots of sin survive after the new birth." Already the colonists had managed to get for themselves all the comforts they needed, both for body and soul. "I tell you the truth that no one can describe a land which flows with as much milk and honey as this one does." [49]

They had reason to be hopeful in those early days, and they made a good impression on outsiders. A Chicago correspondent for *Harbinger*, published by the Brooks Farm Colony at Roxbury,

Massachusetts, said that he had seen about sixty-five of them camping outside his window in 1846, before they set out for Bishop Hill. He said, "There was a look about these people which I have never seen among the masses of European immigrants who have passed through Chicago since I have lived here. It was an expression of patient, intelligent endurance; all had it, except the young children. . . . . They walked erect and firm, looking always hopeful and contented, though very serious." [50]

They were also beginning to feel more and more at home in the new land, learning enough of the language to take a modest interest in the various problems agitating the American scene at the time. They cannot be said to have been deeply involved, not even in the slavery issue, which was being debated hotly on all sides. George Washington Gale's utopian colony at nearby Galesburg was taking the leadership in the antislavery fight in the Fourth District, and Lars Paul Esbjörn was violently opposed to slavery at nearby Andover, but the Bishop Hill utopians were not much concerned. Von Schneidau wrote from his daguerreotype studio in Chicago to Eric Jansson and reminded him that the old law, which required five years of residence before an immigrant could vote, had been changed, and now the newcomer could vote after six months. He urged Jansson to get his colony's support for the candidacy of the Whig, Zachary Taylor, in the 1848 election.[51]

The issues in the 1848 campaign were not very clear. Only the Free-Soil party took a strong line against slavery, and both the Whig Taylor and his Democratic opponent Lewis Cass pussyfooted about the matter. Taylor was somewhat more acceptable to the free-soil states. He was a large slaveholder himself, but he said he would call out the troops if anyone tried to prevent the entry of two free-soil states, California and New Mexico, into the Union. Cass went about saying weakly that every territory ought to be allowed to decide for itself whether or not it would make slavery legal, and Eric Jansson supported him. Von Schneidau's pleas fell on deaf ears, since the Bishop Hill men were Democrats, as were almost all foreigners at the time; and Captain Johnson recalled seeing a column of 120 colonists marching four miles through two feet of snow to cast all their votes for Cass.[52] But Zachary Talor was elected by a comfortable margin. The colonists would have to wait another thirteen years, until the outbreak of the Civil War, before another column would march off from the village, this time in support of the northern cause and the right of black men to come into the same heritage of freedom the Janssonists had sought.

Portents of Disaster

Mars, du hast es besser.

—Erich Bloch, *Das Prinzip Hoffnung* (1956)

There were ominous signs that the colony dream would be difficult to realize as soon as the harrowing ocean voyages began, and disillusioned Janssonists began drifting away on their own when they reached New York. [1] But the solid core of believers clung to their faith, regardless of the negative evidence that began piling up. There were predictable difficulties involved in trying to get along with people whose language they did not understand. There were obvious difficulties in setting up even a small city on the prairies of Illinois, with the nearest shopping center of any size at Rock Island, fifteen miles away. And there were increasing troubles which sprang up amongst the colonists themselves.

Eric Ulric Norberg of Ullervad, Västergötland, was one of the very first Swedes to come to the middle West, having arrived in 1842.[2] At first he had settled down at Pine Lake, Wisconsin, near Unonius, but soon he moved north to Pere Marquette, Michigan. On one of his trips south to Chicago, he had met John Edward Lilljeholm, a roistering adventurer newly arrived from Sweden, and invited Lilljeholm to join him in a lumbering enterprise in the northern woods. After several seasons, Norberg loaded their collection of shingles on a boat and crossed Lake Michigan with the intention of selling them at a trading post in Milwaukee. Lilljeholm waited as patiently as Tristram for the return of his friend, but Norberg never returned with the profits.

Muttering uncomplimentary Swedish words, Lilljeholm set off to find his partner. The Swedes in Chicago said that Norberg had indeed been there, but he had left to join the Janssonists at Bishop Hill. And at Bishop Hill, Lilljeholm was told that Norberg had just been there, but had left for parts unknown. One day as Lilljeholm was sitting disconsolately beside an open window, he overheard two of the colonists talking about Norberg, and found out that his partner was hiding at Red Oak Grove, three miles away. He hurried there and confronted his former friend.

Angrily denouncing Norberg and his poor sense of contractual obligation, Lilljeholm turned the air blue, whereupon Norberg suggested that they present the whole matter to Eric Jansson for arbitration. When they came before Eric Jansson, Lilljeholm turned his wrath against the prophet himself. Jansson listened while Lilljeholm described him as a cheat and a swindler, and not only that, but a murderer as well, since he had caused the death of several hundred Swedes. When he was through, Jansson quietly gave him a letter of conveyance for all Norberg's property in Michigan (a team of oxen and some lumber), and as a final gesture gave Lilljeholm twenty dollars to pay for his return trip.[3]

Norberg figured in another confrontation with Jansson, this time on the plaintiff's side. It is likely that Jansson told Norberg exactly what he thought of Norberg's conduct in the Lilljeholm affair. In any case, there was growing dissension between Jansson and Norberg, and with the rest of the colony as well, though no one had the nerve to tell Jansson so to his face. But Norberg dared. He chose to talk to Jansson the day after his wedding. Some foolish business transactions had plunged the colony in debt. Unonius tells us what happened: "Dissatisfaction ran rampant among the people, but Erik Jansson had known how to imbue them with such fear that no one dared openly oppose him. One Norberg, a former Swedish bailiff, was the only one who, at the urgent appeal of the Olssons . . . dared to come to him with complaints and reproach him for his mismanagement and his arbitrary actions in general. To this Jansson replied with his usual 'I have acted consistently with my preaching. Those who are dissatisfied are deceived by the devil.' "[4] Bedeviled or not, Norberg moved out at once, and on March 23, 1850, he was listed as one of the members of the Evangelical Lutheran Church in Andover.[5]

Another one of the local problems curiously anticipated the Root case. Olof Bäck, who was one of the large group from Bollnäs, wrote a long, denunciatory letter to Sweden on December 27, 1848. He said that he had come over to America only because of his wife's insistence, and had never been in his heart a Janssonist.[6] Bäck said that he had tried to rescue his wife from the "slave camp" of Bishop Hill, but he had been unable to do so.

"The Janssonists," he said, "stole her and everything she had from me, though she never intended to stay there. And so I must live alone like a hermit." [7]

Bäck took his revenge on Jansson by writing scathing letters back home, filled with unsubstantiated charges against the Janssonists. The editor of *Norrlandsposten* in Gävle published six of these letters in 1848 and 1849. On March 31, 1849, for example, Bäck reported that he had visited the colony and found the settlers in deep trouble. When he had been there in the fall of 1848, things had seemed quite different: they were shipping carloads of wheat to Chicago and Peoria, and were renting out to neighboring farms a threshing machine so large that it took two wagons to transport it. But the spring of 1849 was another story. Cholera had struck the colony, and a great many had died. As soon as some one fell ill, Bäck said, Jansson went to the bedside and accused him of falling away from the faith. "And in that way . . . he drives them to work until they fall over dead. Then they get his last word: 'You shall be trampled down into the stones of hell because of your disbelief!' " Only four hundred were left of the original twelve hundred emigrants, the rest having either died or deserted to such places as Galesburg, Lafayette, Victoria, and Andover.[8]

Bäck had a personal vendetta to carry on, but he was right that cholera was a problem which left no one in the colony untouched. It destroyed for all time the optimism of the first two years. First breaking out in Illinois in 1832, during the Black Hawk wars, cholera had been especially virulent around Rock Island, where the military garrison was decimated, and it had also been of plague proportions downriver at St. Louis, which had seemed that summer like a vast charnel house.[9] But the epidemic had mysteriously subsided and the disease had lain dormant for a decade, only to break out in full fury during the summers between 1848 and 1854. Cases were reported from New York to New Orleans, but it was obvious that the channels of transmission were the pioneer trails to the west, especially on the waterways. The lake boats were death traps, as were also the Mississippi riverboats, and St. Louis seemed the focal point of the plague.[10] The full force of the epidemic was felt in April and May, 1849, when it is estimated that somewhere between four thousand and six thousand people died in the Middle West.

The convergence of several factors brought on the epidemic: the presence of the germ itself, of course; the crowded quarters and poor sanitation on the overland trails and on the waterways; the fatigue of the travelers; their inadequate diet; and the near absence of qualified medical aid.[11] The symptoms were easily recognizable: a progression through mild diarrhea, such as Esbjörn suffered when he arrived in Chicago in the fall of 1849,[12] to prolonged vomiting, physical collapse, uremia, coma, and death. Many experienced only the early stages, and as they did not know how they had become ill, became well just as mysteriously. But great numbers died with surprising rapidity, often feeling well and strong in the evening and in the morning lying dead.

Both those who escaped the infection and those who knew it in mild form were terrified. "We also may be marked as victims," a reporter for the *Chicago Daily Democrat* speculated on January 17, 1850, "and the shaft of death be already on the poise and aimed at our hearts." Sometimes there was the decimation of whole villages, and then there was panic flight, desertion of family and friends, leaving the sick untended and the dead unburied, or wrapped hastily in a blanket and dropped into a common grave. John Lilljeholm said he saw Mississippi riverboats beached on mud flats because the crew was too sick to sail them, and he himself fled headlong through deep mud when he saw his fellow cabin mate lying dead on his bed.[13]

Cholera came to Bishop Hill in July, 1849, with a party of Norwegian converts who had been persuaded to take the journey by Jonas Nylund. They had picked up the germ somewhere along the route, and many were sick when they arrived. Ironically, none of the Norwegians died, and in a few days all except three of them left Bishop Hill, some becoming Mormons, and some scattering through the Norwegian settlements in Illinois and Wisconsin.[14] Nightmare followed for the colony from July 22 to the middle of September. Jansson ordered those who were well to flee to the colony farm at La Grange, Illinois (now Orion), where they soon filled up the available buildings and lay in tents under the trees. But now they were even closer to the center of infection, at nearby Rock Island, and death stalked through their encampment.[15] Immune because of physiological accident, or because of saintliness, or for whatever reason, Eric Linden went about and cared for the sick, and when they died, buried them without a coffin in a common grave. Jansson himself took his family to what he took to be a sure refuge, the colony fishing shacks on Arsenal Island near Davenport.[16] But the angel of death was not to be outwitted, and before the holocaust was over, Maja Stina, his long-suffering wife from Klockaregården, lay dead.

It is indescribably sad to read about the deaths of these victims of cholera, since methods of treatment in the nineteenth century—when they were used at all—had the opposite effect of making death almost certain. Those who were ill begged for water. Dehydration set in when the body fluids drained away, and then the kidneys could not deal with the excess toxic substance. Treatment at that time consisted of denying the patients water, and giving them large doses of calomel, camphor, red pepper, and mustard footbaths.[17] Professor Robert Koch isolated the cholera organism in 1883, and treatment now is the relatively simple matter of replacing lost body fluids.

The cholera epidemic was catastrophic for the Bishop Hill Colony, not only because two hundred of the members died, but also because it exposed once and for all the theological weakness of the perfectionist principle. According to the party line, those who had received the new birth and had become one with Christ had nothing to fear from such blandishments of the devil as bodily illness. The theory could still be maintained when an occasional death occurred, but when scores were carried away, including leaders of the colony such as Olof Olsson, and members of Eric Jansson's own family, it became much less convincing to claim that death was a penalty for disbelief. Jansson persisted in saying that very thing, denying any role at all to secondary causes such as infection and unsanitary living quarters; but it became more and more plain that something other than a theological decision was called for, and that medical help was necessary. Jonas Hedström came over twice a week from Victoria, and moved about sick colonists, sometimes taking a wagonload of them home to die on his kitchen floor.[18] Finally some American neighbors and he confronted Eric Jansson and threatened that if a doctor were not summoned at once, they would report the matter to the authorities and hold Jansson responsible for needless deaths. Jansson agreed to bring in medical help.[19]

The doctor who was brought in was an American quack, Dr. Robert D. Foster, who promised blithely that he would save ninety-nine of a hundred cholera patients. Unfortunately, he was a poor physician but a shrewd businessman. Unonius was not an unprejudiced witness, but he probably was right when he described Dr. Foster as a charlatan: "Though he knew very little of the healing art, he knew all the more how to gain the unlimited confidence of Jansson, whom he swindled out of considerable amounts, bringing him and the entire community almost to the brink of ruin." [20] He induced Jansson to make some rash investments, he sold him a parcel of land at an inflated price, and he charged him gouging fees for even trying to help the cholera victims. For much of this, Jansson could only give him promissory notes. As security on these notes, Jansson signed a mortgage on all the chattels and inventories of the colony. When he could not pay, Jansson forfeited nearly everything the colony owned to the doctor, who held a public auction of his take. The Janssonists had to stand helplessly by in the fall of 1849 while the doctor auctioned off thirty pairs of their oxen, eight or nine pairs of horses, ninety-four calves, hogs, wagons, farm tools, grain and food supplies, and even some used bedding.[21]

The financial expense of the cholera epidemic was staggering, but this was recouped in the decade that followed. What was not recouped was the loss of so many trusted and loved friends. In September the shocked survivors began to understand the depth of their loss. Olof Olsson was one of the first to fall, as had also his wife Beata and their two children.[22] Lars Larsson from Hov, Söderala Parish, was gone; so was his wife, Anna, and four of their children. Also missing were Nils Pehrsson of Bers, his wife, Anna, and his son, Pehr; only his daughter Margta was still alive and needing a home. Carin Johnsdotter of Norrby was left a widow when her husband, Olof, and their only child died. The paper-mill owner from Forsa, Jonas Lundquist, and his sister, Anna, were both dead. So was Erich Lars Olof, from Norrby, and Sigrid from Heden. Pehr Anderson Lund, his wife Kerstin, and the youngest of their four children died. Nils Olofsson from Sunnanå died, though his wife Kerstin and their four children survived. The grenadier, Johan Hård, and the spinster from Söderhamn, Charlotte Frank, were dead.[23] The list went on and on—over two hundred whose lives in the land of milk and honey had come abruptly to an end.

During that heartbreaking summer the colonists had been

numb with fear. Victor Witting, the sturdy sailor who went about caring for the sick, said that death became so commonplace that "one hardly felt it, and paid little attention to who had died." [24] But later on they would gather in the cottages and exchange names. Jansson had ordered that the corpses should be buried at night, without ceremony, so that no one had any clear idea of how many had died, or where they were buried, nor even if they died from cholera or from the prophet's displeasure.[25] He had let the word spread among his credulous followers that he had the keys to life and death, and the simple minded could believe that the price of rebellion against his rule was too high to pay.

Jansson had reason now not to permit any strident questioning of his business management. Daniel Londborg reported that the fact of his incapacity for commercial transactions was much talked about by the survivors of the cholera epidemic. He had saved his personal possessions from Foster's grasp, but he was said to be in debt to the doctor for all he owned, and in addition owed $150,000 to creditors in Chicago.[26] The amount is surely incorrect, but that they all faced financial disaster was clear enough. The purchase of the ten thousand acres from Dr. Foster seemed an especially foolish move. There never was a sturdier band of immigrants who settled the plains of Illinois, but now even their spirits were at a low ebb.

There was no sign of weakness from their leader, or any sign that he doubted his mission. He had improved his appearance though not his speaking ability by investing some of the colony funds in a set of false teeth. Daniel Londborg has given us an interesting account of his visit to Bishop Hill made together with several other dissenters on Sunday, October 21, 1849. He was not especially pious but he decided that the polite thing to do was to go to the Colony Church, where the service had just started. Ushered in on the men's side, they were given hymnbooks and sang one of Jansson's long hymns. Then the prophet began to preach. His theme, according to Londborg, was his usual one: "about his similarity to God." Londborg noticed the trouble he had in articulating words: "Having been justly punished [for his sins] Jansson had lost not only his overflowing front teeth but also the more necessary ones as well, and had gotten a new set of teeth in order to carry on his work as preacher. But despite the new teeth, he slobbered so that I could only with great effort understand what he said. Anyway, I had no great desire to listen any more to his absurd sermon, and so started to walk out, intending to look over their living quarters. Whereupon Jansson, who knew very well that I despised his teaching, broke off what he was saying and called out, 'There you see with your own eyes that the devil has to leave, because he can't stand to hear God's word preached in its purity!' "[27] Londborg continued out and clumped down the stairs.

In order to raise money for the payment of the colony debts, two of their most persuasive spokesmen, Olof Jonsson of Stenbo and Nils Hedin, the tailor, were sent to Sweden in the spring of 1849. Though they were not averse to making a few converts as well, their mission was to collect money from the estates of the sixty Janssonists who had drowned on the shipwrecked *Betty Catharina* the summer before. The Janssonists were not held in high esteem in Sweden, having decimated churches and broken up families, and the returned missionaries were not welcomed. The editor of *Hudikswalls Veckotidning* warned his readers that the agents were in the area, and told them to hold on both to their money and to their souls.[28] But the agents did manage to collect some of the money due their fallen friends, and returned to Bishop Hill around Christmas time with six thousand dollars.

Some hope was raised by the return of "Stenbo Olle" and Hedin with the Swedish money, but the wounds of the colony were too deep to be so easily healed. The desertions from the colony were heavy in the fall of 1849. Such stalwarts as "Fru Hebbe," who had been one of the truly enthusiastic followers and had instructed illiterates during the first winter, decided she had stood enough and left with her family for Galesburg. Mina Skoglund married one of Eric Jansson's servants, Pehr Dalgren, and the bridal pair moved at once to Victoria.[29] The Janssonist dissidents helped to found the colony at Galva, and former colonists were to be found all over Illinois. Londborg said that it was mainly the Söderala contingent who remained loyal to Jansson.

By the spring of 1850, morale at Bishop Hill was at its nadir. L. P. Esbjörn, watching the decline of the Separatists from Andover, wrote an acerbic letter on May 23 to the state church temperance minister, Peter Wieselgren: "I mentioned Eric Jansson. His colony at Bishop Hill is both spiritually and secularly speaking in awful shape. Perfectionism has fallen so low that the brandy bottle and church drinking parties have recovered their prestige, and a distillery is being built by them. Their business is chaotic and the colony has large debts. Also the doctor who was there during the cholera plague last summer has taken many horses and animals for his fee. Also Jansson's artificial gold teeth, which he has had installed to take the place of his own swine-like tusks, and many other such projects, have cost enormous sums." [30]

After Maja Stina died at Rock Island, Jansson returned at once to the colony to find that a rebellion was brewing. As usual in such circumstances, Jansson immediately took the offensive, putting his critics into a defensive posture before they knew what was happening. He faced his congregation in the Colony Church with angry words. His own wife, he said, the spiritual mother of the colony, had become a victim of the cholera epidemic. And why? Because of the "satanic unbelief" of the people. It was only a few days since she had left them, but the colony would not be forgiven unless he married again: "Israel could not be spared unless he remarried forthwith, in order that the spirit and power of the spiritual mother might be present and beautify Israel with its blessed fruits. . . . After the havoc wrought by the angel of death, the voice of the bride and bridegroom was to be heard in the streets of Jerusalem; no weeping or wailing over lost mates and children was to be heard, but everywhere there was to be joy in the Lord."

He said that he did not yet know who was to be the new spiritual mother of the colony, but he had received a revelation that the woman who was to be his bride would precede him that evening into his sleeping chamber. But when evening came, he found not one woman with a distinct call, but two, each one convinced that she was to be the bride, and each one denouncing the pretensions of the other. Jansson had to make a decision, and the woman he chose was Anna Sophia Gabrielsson.

The decision could have been no surprise to the colonists. Sophia had been close to the Janssons since they had arrived in New York in June of 1846. Her second husband, Mr. Pollock, had gone along on the pioneering trek with great reluctance, and he had soon died on the uncivilized prairie; some say because his heart was broken. On July 9, 1848, Sophia married for the third time, this time to the young son of "Guld Gubben," Lindjo Lars Gabrielsson from Dalarna. But the bridegroom's joy was short-lived. He died very soon afterward of Asiatic cholera, and Sophia took out her widow's weeds for the third time. But she looked well in whatever she wore, and Helena Lindewall remembers her as dressing better than the other ladies.[31] She practically joined the Jansson household, mothering her favorite Eric, and nursing Maja Stina during her last illness at Rock Island.

So it was no great surprise to the colony that Sophia was now to become their spiritual mother, and they came dutifully to the Jansson home on September 16, 1849, where Jonas Olsson officiated at the marriage ceremony. We have no record as to the mood of Sophia, who was henceforth known as "Biskopinnan Jansson," but we know that the bridegroom was in high spirits, laughing and joking. The other guests tried very hard to be jovial, but there was a curious atmosphere of depression at the festivities. Jansson himself wondered what was wrong, and said afterward that the affair was more like a funeral than a wedding feast.[32]

11

The Shooting

ERIC JANSSON: Did the world stand still that day in May? I never knew.
SOPHIA JANSSON: No, Eric. There was only darkness for a little while.

—George Swank, "Voices in Retrospect"

The payments to Dr. Foster had all but ruined the colonists financially, but they still had their lands and their buildings as well as their unbroken desire to build a holy city in Henry County. In 1850 they owned four thousand acres of land, an attractive church building, a four-story dwelling house, two other brick houses, five frame buildings, a gristmill and a flour mill run by steam. Altogether they were said to be worth fifty thousand dollars. Though approximately 200 of them had died from cholera, and another 350 had either died on the way over or left after they arrived in America, there were still 550 of them left—100 men, 250 women, and 200 children.[1]

Among the women of the colony was a cousin of Eric Jansson, one Charlotta Lovisa, known by everyone as "Lotta." She was born in Österunda in 1824, the daughter of Anna Mattsdotter and Jon Andersson. Her father had died even before her birth, and Anna Mattsdotter decided to join the Janssonists and travel with her three daughters to America. Anna was the sister of Johannes Mattson, who was the father of Eric Jansson. So Lotta and Eric were cousins, and thereby hangs a tale.

The ship on which they sailed for America was the *Charlotte*, and aboard were many Västmanland neighbors, including two of Eric's brothers, Jan and Pehr. This was before the day of women's liberation, and according to the patriarchal conventions of the time, the oldest male relative served as guide and protector for the women and children in the family. That made Jan the father figure for his Aunt Anna and his cousins, but when they reached Chicago, he left the Janssonists, as did also Carolina,

Lotta's older sister. The rest of the party, including Anna and her two other daughters, Lotta and Sabina Ulrika, continued on to Bishop Hill. Anna died soon after their arrival there.[2] It is worth noting for our story that, according to traditional practice, Eric was then given the charge of his two cousins.

Among the men were three strangers who had drifted up from the southland with fascinating stories of derring-do. Their names were Erik Wester, Charles Zimmerman, and John Root, and they had arrived in the autumn of 1848 from New Orleans. Adventurers in the style of the Duke and the King in Mark Twain's *Huckleberry Finn*, they had moved through the disreputable society of the river towns and on the boats which plied the Mississippi, and they found the virtuous and gullible Swedes of Bishop Hill easy marks. Wester, who had been baptized Westergren in Sweden, had left his native land in great haste. Serving as purchasing agent for Svenska Riksbanken, he had gone down to Helsingör to buy a supply of fine paper which the bank intended to use for notes and documents. But when he reached Helsingör, Wester decided that it would be a far better plan to continue on to Copenhagen and from there sail for America, financing his journey with the funds stolen from his employer.[3]

His first stopping place in America was at New Orleans, where presumably he met two countrymen with equally flexible moral standards, Charles Zimmerman and John Root. The three drifted northward toward Illinois, telling stories along the way about having had important functions in the Swedish army, of having fought with the French Foreign Legion, and most recently with the United States Army in the Mexican War.[4] They came to Bishop Hill looking for wives. Charles Zimmerman moved on to California, when stories of fantastic strikes began coming back from placer miners; but Wester and Root decided to stay in Bishop Hill and sift whatever gold was to be found in the local streams.

Somewhere along the way Wester had picked up some skill as a barber, and Jansson took him on as his personal servant. When the marriage taboo was removed and couples began parading to the windmill every Sunday, Jansson ordered him to marry one of the colony girls who had as yet no suitor. Wester refused. There were rumors that the reason why Wester could not marry the girl was that he already had a wife in Sweden;[5] but Jansson did not

Charlotte ("Lotta") Root (1825–1905), born in Österunda, Västmanland. Her mother was a sister to Johannes Mattson, Eric Jansson's father, so she and Eric were cousins. She became the wife of John Root, and so precipitated the quarrel which led to the shooting of Eric Jansson. Painting by Olof Krans, courtesy of the Bishop Hill Heritage Association

know this and clapped Wester into jail for insubordination. Wester spent no time moping in jail, but began preaching eloquently to his fellow prisoners. He impressed everyone with his Pauline gift, with the result that he was released from prison and invited to preach in the Colony Church. Soon tiring of his new role, he packed up his barber's tools and left for Galesburg, followed by a small flock who now acknowledged him to be their spiritual leader. But when he came to Galesburg he turned on his followers and told them to go to Hälsingland.[6] From then on he carved out a vaguely disreputable career as a small merchant.

The third member of our trio, John Root, has baffled even a very determined effort to uncover his Swedish past, probably because he took a new name when he came to America.[7] Cultivated in manner and impressive in appearance, like Wester he spread the story of having been born to a rich Stockholm family and educated for an army career. He told his credulous listeners that he had been in the Swedish army and had fought with the United States Army in the Mexican War.[8] Later on, after he had been thoroughly discredited, there was a story circulating that he had left Sweden precipitously, a fugitive like his friend Wester from Swedish justice.[9] But when he arrived at Bishop Hill he seemed all that he said he was: handsome, debonair, religious, who could be for them what Polycarpus von Schneidau was to the Chicago Swedish community, a cultivated gentleman with a military bearing and an exotic past.

Whether he was moved by prudential or by romantic motives, John Root began courting Eric Jansson's cousin, Lotta Jansson, and in November, 1849, the two were married.[10] There is disagreement about the marriage contract. John Root and his friends said later on that there was no contract at all, but Lotta swore that he had agreed never to force her to go with him if he left Bishop Hill. Apparently at her very wedding, Lotta had some apprehension about her husband's trustworthiness, because she said he signed a document with this sentence: "If it should happen that John Root should lose his faith and wish to leave the colony, I as his wife have complete right to stay with my friends and relatives as long as I wish, without any interference from him." [11]

John Root quickly justified Lotta's misgivings about him as a husband. He showed up for the common meals, and drew his proper allotment of clothes from John Hällsén, the tailor. But he was never known to share in the work of the colony and was most often seen to be swinging off into the woods with his hunting rifle on his shoulder and his bowie knife at his belt. For several months after he was married he was not seen at all. In one of the first serious studies of the colony, Professor Mikkelsen said that he had taken a job as guide and interpreter for a Jewish peddler who sold his wares in the territory, and that after a time the peddler had disappeared. Some years later, the colonists found the decomposed body of an unknown man under the floor of an isolated cabin, and there were those who nodded their heads and said that Root had murdered him.[12]

There is no evidence at all for Mikkelsen's theory, but there is plenty of evidence which explains why Root was suspected of such a crime. Yet in the whimsical way in which historical data sometimes sifts down to later inquirers, Root emerges with a curiously ambiguous character. On the one hand he gave the impression of a gentleman putting up with frontier gaucheries, and seemed colorful and urbane, friendly with the neighboring Americans, and ready to swap a yarn or two over a glass of beer. Later on when the troubles broke out, State's Attorney Harmon G. Reynolds wrote an official report of the whole affair for the information of Governor French, and he took pains to speak well of Root: "He was," said Reynolds, "an educated Swede, and a gentleman in his manners and intercourse in Society, and was mustered in luxury and indolence." [13] But he was also a scalawag. Captain Johnson, whose father was to become his victim, said of him that he was "the worst scoundrel who ever set foot on American soil." [14] And Pastor Unonius, who must be counted a neutral observer, said that Root was "as sly and ungodly a rascal as Jansson himself." [15]

It was not long before he was in trouble at Bishop Hill. One quarrel was with the doctor, Robert Foster, and the argument grew so heated that the colonists agreed that one of the two disputants must leave Bishop Hill. To the dismay of Eric Jansson, they voted that Dr. Foster must go, and the Swede, John Root, could stay.[16] Root's next quarrel was with Eric Jansson himself. Root had come into some money, and Jansson, of course, expected him to contribute it to the general fund.[17] But Root had no such communal loyalties and refused to part with his dollars.

As the senior male in the relation, Jansson was also involved with Root's basic quarrel, which was with his wife. Root had come back from one of his long absences and had discovered that his wife had borne him a son.[18] The colony was suffering the cholera scourge, and Root thought that they should flee for their lives. Having a Swedish notion of the man's role in the family, he ordered his wife to pack up and be ready to move out with their son. He was shaken to hear her refuse.

The grounds of her refusal were much debated in the years which lay ahead, and perhaps no single reason is adequate. She was a very pious and domestic sort of young lady, and she felt safe in her home surrounded by other believers, raising her voice in song at the Colony Church and joining her friends at the common table and the common work. On the rare occasions when he was home, Root was roughly masculine and overbearing, sometimes threatening her and perhaps subjecting her to physical abuse. State Attorney Reynolds speculated about her motives in the letter he wrote to Governor French: "Some of the Americans say that she wanted to live with him at Bishop Hill—not elsewhere—others that she was far advanced in pregnancy, could not speak English, and did not want to go for that reason—others again, that she was only restrained by the doctrine taught her by Jansson, that if she or any of them left, they would be damned." [19]

Though some observers of the drama were wrong on the point, Root knew well enough that Lotta did not want to leave Bishop Hill; but he attributed her desire to Jansson's hypnotic power over her. An anonymous letter to the editor of the *Gem of the Prairie* seems to come close to the truth: "He thought that if she could be removed from under his [Jansson's] influence for a time, to a clearer atmosphere, where her mind could regain its natural balance, she would be perfectly satisfied and happy to live with him." [20] But perhaps the simple truth is that Root was afraid that if they stayed in Bishop Hill, they would all die of cholera.

On Saturday, March 2, 1850, a young man from nearby Cambridge, Daniel Stanley, agreed to go with Root to Bishop Hill and force his wife and child to come away. Piecing together some eyewitness accounts, we know that Root ordered her to go with him and that she refused, whereupon Root said that in that case he would take their son away. Stanley started down the steps from their rooms on the second floor of Big Brick carrying the baby in his arms, and Lotta hastily gathered up some clothes and followed. With Root driving the horses and with Lotta and the baby and Daniel Stanley in the back, they raced out of the village. The colonists were eating in the basement of Big Brick and knew nothing of all this until a spectator rushed down to tell them. Twelve of the men hurried out, saddled horses, and raced out in pursuit. A few miles outside the village they intercepted the carriage, and asked Lotta if she wanted to go with her husband or stay in Bishop Hill. She rose up to reply, whereupon Root laid down his pistol to pull her down. At that moment one of the colonists (he is said to have been Jacob Jacobson) dove at Root's pistol and wrenched it away. Stanley saw that the jig was up and turned over his pistol. Lotta returned with her friends to Bishop Hill.

The next day Root turned to legal means to recover his family, and swore out a writ for the arrest of Jansson and others, charging them with riot. Lotta Root was subpoenaed as a witness and went to Cambridge as a witness. When she got there the clerk of court, S. P. Brainard, held her incommunicado where the colonists could not speak to her. Since Root himself failed to show up to press charges, the case was dismissed, but shortly afterward he arrived and took his wife and child to Rock Island, where they stayed with an American friend, P. K. Hanna. After a few days they left for Chicago and the home of Lotta's sister, Caroline, who had married Pehr Ersson. The Erssons were charter members of St. Ansgarius Church, and lived in a neat house near the church, just north of the Chicago River.[21] Caroline sent word to her cousin, Eric's brother Jan, who lived nearby. He waited until John Root had left the house one day and quickly rescued Lotta and the child, taking them back to Bishop Hill.[22]

Root was beside himself with rage when he found out that for the second time Eric Jansson had managed to take his wife from him. The neighbors said he acted strangely: he was sometimes silent, and at other times he would rage incoherently, banging a newspaper against a table, brandishing a weapon, and speaking of vengeance. He began practicing firing his pistol. To anyone who would listen he said that Jansson had stolen his wife and child and had blackened his reputation. Root toyed with the idea of suicide, and told people that he would be better off dead.[23]

Many of the Lutheran and Episcopalian Swedes of Chicago shrugged their shoulders, but some were sympathetic and spoke of his right to redress. When Root told his tale in Henry County, he won some more believers, who were willing to believe any charge against these arrogant and pious Swedes. Violence was just below the surface in the Illinois of 1850. It must be recalled that the state was newly fledged, the roads were rutted lanes, the structures of justice and law enforcement were primitive and ineffective. Bands of outlaws sometimes roamed the state, striking terror in the more civilized citizens. The bandit gang of Cave-in-Rock had been broken up, but as late as 1837 a mob held Pope and Massac counties in its grip, ruling from a private fortress. In 1837 a band of hoodlums threw the press of the *Alton Observer* into the river, and then killed its editor, Elijah Lovejoy. Murders, robberies, horse-stealing, and counterfeiting were commonplace. When an Ogle County family of criminals named Driscoll shot down Captain Campbell, a kangaroo court heard the case and found Driscoll and his son guilty. The two were made to kneel, and one hundred citizens fired at them in unison. The executioners were tried and acquitted. On June 7, 1844, the year before the Janssonists began arriving, a mob attacked the nearby Mormon colony at Carthage, and killed its leader Joseph Smith and his brother Hyrum.[24] There were grounds for the belief that Bishop Hill would suffer the same violence.

John Root managed to focus some of the vague tendency toward violence in the Illinois countryside. The Americans at Walnut Grove and at Red Oak Grove were suspicious of him, but he had friends at Cambridge, Geneseo, and at Rock River. At Cambridge, especially, they sympathized with him, since he was a member of the local Masonic lodge. On March 26, Root gathered such a mob of vandals together and led them to Bishop Hill to recover his wife and child. She was not there. Furious, Root incited his followers to blind rage, and they marched up and down the streets of the village, brandishing their weapons, searching the houses, and promising to burn the whole place down if Lotta Root were not produced in eight days. Nils Nilsson from Källeräng, Delsbo Parish, tried to hide in the woods, but he was found and threatened with hanging. John Root interceded for him and his life was spared.[25] After searching without success for Lotta, most of the mob left, but some stayed, and the colonists put them up for the night and gave them food to eat.[26]

The next night, on March 27, they came back again, this time tearing off some boards from the Colony Church and houses. They fired guns into the air and ordered the inhabitants out of their houses, intending as they said, to burn the town to the ground. The men were herded into the basement of the church, and the women and the children into the hospital. After finding that the colonists offered no resistance to their outrageous commands, they seemed suddenly ashamed, let the people return to their homes, and rode away. But on the following day another mob came, and this time they set fire to some haystacks and burned some buildings at Little Hill, about two miles west of Cambridge.

The mob seemed to be more interested in bringing off their shenanigans than in proving the justice of their cause. One band of drunken revelers came down the road in a wildly careening carriage. When Magistrate Piatt confronted them and told them that any display of vandalism or terrorism would land the offenders in jail, they were just sober enough to drive off. Besides, John Root himself was no longer on the scene, and the sympathy of the mob began to turn against him. If Root were not interested enough to be present, why should they press his cause?

Jansson also could not be found because he was hiding under the floor of a farming outpost south of the village called "Sör Stuga." The situation was much like that in Sweden, when he was fleeing for his life, except that now it was his neighbors rather than the police who were in hot pursuit. When Jacob Jacobsson arrived, it was decided that he should continue to Alton, ostensibly to buy flour, but really to escape the desperadoes. An old woman was sent up to the village to reconnoiter and to bring back a report to the refugees at Sör Stuga. She told them what was going on, and they decided to flee south to St. Louis. Together with Jansson and Sophia, his bride of a few months, went Lotta Root and her son, Jonas Olsson, Peter Johnson, Erik Stålberg, Olof Lind, Peter Nelson, and Per Olof.

It seemed sensible to let Jonas Olsson go with a party from the colony who were intent on searching for gold, since they certainly would escape the fury of John Root, and might even bring back enough gold to pay off the colony debts. The plan was that Jonas Olsson and five others would go with the St. Louis refugees, and buy supplies there, while another party of three would strike out westward across Iowa, and rendezvous at Fort Kearney, Nebraska. The gold expedition proved a failure. Jonas Olsson kept a diary of his daily adventures, and we have a detailed account of their hardships, frustrations, and indifferent success.[27]

Meanwhile in St. Louis, Eric Jansson reverted to his original role as a flour salesman and exulted in the ready market which the big city offered for whatever wares they had for sale. Two letters from him to the colony have survived from this period, and they show him to be by turns terrified at the threat to his life and to the village, discouraged about the heavy debts hanging over them, and at the same time excited about St. Louis as a territory for sales. On April 17, he wrote to Andrew Berglund, who was the head of the community when Jansson himself and Jonas Olsson were not on the scene.[28] "We have fled," he wrote, "from our murderers, and we have heard how Root and his mob have threatened to destroy our homes and our buildings." The colonists were obviously in deep trouble, but they were not to be discouraged. What was happening was simply what always happened: the forces of darkness were trying to overwhelm the children of light.

He then turned quickly to commercial concerns. It is true that they had been bilked in their dealings with Robert Foster, having paid three thousand dollars for property which they were now able to mortgage to Hall and McNeely of St. Louis for only two thousand dollars.[29] But that money would come in handy and could be used to finish construction on the colony's steam-driven flour mill. After strenuous effort, Jansson was finally able to buy flax seed, which was in short supply, and he had bought it at a favorable price. He had also managed to buy enough iron to produce at least thirty wagons. The colonists should not rest from their labors. Parties should be sent out to harvest slippery elm bark, which could be ground up and sold in half barrels. The women should stick to their weaving: he knew hungry customers when he saw them, and he thought three thousand mats could be sold in St. Louis.

They need not worry about his personal safety, he assured Andrew Berglund. He had been threatened many times in his life, but had always passed through unscathed. "It is true as I have said that Rut [Root] is after my life. But the Lord will not let me fall into his hands." He felt a sneaking respect for Root, who had received so much at his hands and now was causing such dire peril. "I have said that no one has tormented me so much as

Rut. But he is on that very account a devil and will have hell as payment for all his spunk."

Three days later, on April 20, Jansson wrote again to the colony, this time making a desperate attempt to calm their fears of business disaster.[30] Word had come to him that the colonists were worried about business affairs, some having claimed that their outstanding debts were so large that they exceeded the total value of all the property they owned. They should take heart he said; no bankruptcy was in the offing. A way out of the financial plight now presented itself. They would simply not pay their debts owed to worldlings, and they would redouble their efforts to produce saleable goods.

The biblical authority he invoked was from Exod. 3:18–22, the story of how the Israelites had "borrowed" silver and gold jewels from their Egyptian captors before setting out on the Exodus. "They knew they would never pay it back," said Jansson. "So far I have never gone, but if God forces me to be like my fathers, how can I help it?" The case was perfectly clear: the world wanted to destroy the Israelites, and the Israelites were allowed to take any steps possible to avoid destruction. "Do not," he warned in a realistic mode unlike his normal posture, "be so righteous that you destroy yourselves!" Root had begun to appear in his mind as Pharoah, leading the savage Egyptians against God's chosen people: "Is not Rut like Pharoah, who tried to prevent God's people from worshipping their Creator after their own conscience? And does he not in a thousand ways torment God's people? Should not God also be allowed deception as a way out? Can you honestly say that Israel's folk have ever gotten anything from the world, except by cunning and theft?" [31] The years of disguises, of lying, of flight from capture, of half-truths and denials had begun to take their toll.

While Jansson and his wife were in hiding in St. Louis, his son Eric had been sent to Toulon, Illinois, where he was given refuge by W. W. Drummond, a friendly attorney. The lawyer assured Sophia of her stepson's safety, and said that the boy was well cared for and had in fact been trying to teach them Swedish. Drummond's concern for Jansson and his family was moving. He had heard that the mob decided to wait only eight days for the return of Lotta Root, after which they had determined to burn down the village.[32]

Another lawyer helped the refugees in St. Louis. They had

sought out Britton A. Hill and had received his professional aid in drawing up documents intended for the governor of Illinois, Augustus C. French, in the hopes that he would use the forces of legal order to prevent the mob from carrying out its will. Lotta swore out an affidavit which was intended to refute once and for all the charge that the colonists had kidnapped her against her will.

> April 8, 1850
>
> State of Missouri
> County of St. Louis ss:
>
> Charlotte Louise Root wife of John Root, being duly sworn says, that about the 18th day of March 1850, she voluntarily left her husband while they were at Chicago and went to the Swedish Colony with her friends. Affiant left her said husband on account of ill treatment and abuse, and she went to the Colony aforesaid of her own will, without being persuaded thereto by any one. On the 22d of March affiant arrived at the Colony, and on the 24th of the same month she left, for a place of safety, on account of her husband having threatened to take her away. Affiant proceeded to the City of St. Louis, where she now is, with her infant child five months old. Affiant verily believes that said John Root, will take her life, if she returns to him, and she is afraid to do so. The people of the Swedish Colony have not had any influence upon affiant to induce her to leave her said husband. And affiant being afraid of her own life, declares that she will not live with her said husband any longer. He is a man of violent temper, & terrible passions & affiant trembles in his presence—and although affiant is very anxious to save the Colonists from harm & would do anything in her power to do so, yet she is not able & dare not go back to her said husband. Before leaving him, affiant bore as long as she could with his violence, abuse & ill treatment, & now she is satisfied, that if she returned her life would be sacrificed.
>
> Chearlata Lovisa Root [*sic*] [33]

Britton Hill sent Governor French Lotta's affidavit, together with a detailed account of the whole disturbance. If the governor did not move, it was not because he was unaware of the colonists' case.

After a stay of three weeks in St. Louis, when the danger from the mob seemed over, Jansson and his friends returned to Bishop Hill. He was notified that he must appear as a defendant in several suits filed at the May term of the Henry County Circuit Court in nearby Cambridge, and he knew exactly the danger that was involved. On Sunday morning, May 12, he picked a Scripture lesson which would serve as his theme for the ordeal that lay ahead: "I am already being offered and the time of my departure is come. I have fought the good fight, I have finished my course. I have kept the faith; henceforth there is laid up for me the crown of righteousness, which the Lord, the righteous Judge, shall give to me at that day; and not to me only, but also to all them that have loved his appearing." There were no dry eyes in the congregation as he administered the bread and wine during Communion, and said, "I say unto you, I shall not drink henceforth of this Fruit of the vine until that day when I drink with you in my Father's kingdom." [34]

Early Monday morning, Richard Mascall, an employee of the colony, called at the Jansson home with a horse and buggy to drive him to Cambridge. Coming down the steps, Jansson said with a grim smile, "Well, Mr. Mascall, will you stop the bullet for me today?" They set out on the spring morning for the ride to the county courthouse at Cambridge, fifteen miles away.[35] It was May 13. Jansson was the defendant for the colony in five cases, and John Root was also there, plaintiff in a trespass case against Capt. Peter W. Wirstrom. After the fifteenth case on the docket had been dealt with, all left the second-floor courtroom for the noon recess except Eric Jansson and the clerk of court, S. P. Brainard. Brainard sat at a desk writing some notes, and Jansson stood at a window watching some children at play in the yard below. John Root came rushing up the stairs shouting, "Eric Jansson!" Jansson turned to confront Root, and they exchanged some words in Swedish which Brainard could not understand. It is clear that Root demanded his wife back, because Jansson replied, "A sow should be good enough for John Root!" Root drew a pistol and fired twice, one shot spinning off Jansson's shoulder and the other entering his heart. Five minutes later he lay dead. Brainard entered in the court records opposite the sixteenth case: "Death of the defendant suggested and case continued." [36]

Waking from the Dream

> The majority of those who now dwell in this Colony . . . are outside of all congregations. That they are highly indifferent with respect to theological dogmas is not surprising.
>
> —Eric Johnson, *Swedes in Illinois* (1880)

John Root stood quietly beside his fallen enemy, the pistol smoking in his hand. He made no effort to resist arrest, saying merely that "he had nothing more to live for, as he had accomplished all that he desired." [1] The grand jury was already in session and that afternoon it heard Case No. 16, *The People v. John Root.* The state's attorney, Harmon G. Reynolds, had hastily drawn up an indictment. It stated that John Root with a pistol worth three dollars had willfully and with malice aforethought caused a bullet to penetrate the left breast of Eric Jansson, who had for that reason languished for five minutes and then died. By so doing, the said John Root had committed an offense against the peace and dignity of the state of Illinois.[2]

The judge was the Honorable William Kellogg. In 1850 he was thirty-six years old and newly elected to the bench of the Tenth Judicial Circuit Court, but he had already begun assembling a reputation as a firm and fair arbiter of justice, outstanding both for his common sense and for his integrity.[3] Five years later, in 1855, his downstate friend Abraham Lincoln was to write and ask him what he thought Lincoln's chances were to be elected senator from Illinois.[4] In 1857, Judge Kellogg was himself elected congressman from the Illinois Fourth District, campaigning as a moderate on the slavery issue.

The day after the shooting, on Tuesday, May 14, Root was arraigned to answer the charges brought against him by the state of Illinois. His attorneys were the Messrs. Peters of Peoria and Julius Manning of Knoxville.[5] Root outlined his proposed defense: Eric Jansson had at various times threatened to kill him.[6] In

English inscription on Eric Jansson's tombstone in Bishop Hill Cemetery. Courtesy of the Bishop Hill Heritage Association

fact, he had as recently as March 3 ordered the kidnapping of his wife and child, and had told the kidnappers to kill Root if he tried to oppose them. The witnesses he expected to call to support his case were Johan Hellsen, who was for the time being in Missouri; Eric Wester, of Freeport, Illinois, who would show that Jansson had at various times predicted Root's early death; Jan Jansson, Eric's brother, who lived in Chicago, and who could testify concerning the kidnapping; and the doctor, Robert D. Foster, who now lived in New York, but who knew the colony people and could testify that Jansson had held Root's wife captive in Missouri for three weeks.[7]

The line of defense which Manning had established was now clear: intense and prolonged provocation by Jansson had caused his client Root to yield to "a sudden and violent impulse," but there was no malice aforethought. Root was acting in self-defense. Just before Root had fired his pistol, Jansson had made a move toward his own pocket, as though reaching for a gun. There were witnesses to all this, but Root had no time to find them or to enlist their support. He pleaded not guilty to the charge of murder and asked for a postponement of the trial until he could make a proper defense. Judge Kellogg granted him the stay, gave him a copy of the indictment as well as a list of witnesses, and ordered him confined in the prison at Toulon, Illinois.

Sick at heart, Jacob Jacobsson and Nils Hedin took the body of their leader back to Bishop Hill. The colony was in shock. They were used to difficult times, to persecution and even to death, and had never known any period in the last few years when they had not known suffering. But this trouble was different—a crushing and perhaps mortal blow to the colony. One last hope flickered faintly: God would not allow His only messenger to perish from the earth. They stood weeping and praying beside the coffin in the basement of Big Brick, but the hands that had blessed them lay folded across his waist, and the lips that had spoken so many truths had obviously been silenced forever.[8]

On the day of the funeral the Colony Church was crowded, since many neighboring American sympathizers and sightseers joined the mourners. There were prayers and a hymn. Then Mrs. Jansson rose from her pew, stepped forward dramatically, and placed her hands on the head of Andrew Berglund, naming him the titular head of the colony until Jansson's own son would be old enough to take the leadership. Berglund preached in English, and there were no other eulogies. When the service was over, they carried the plain wooden coffin in procession to the burial ground, lowered it into a grave, and placed a wooden slab over it to mark the spot.[9]

The drama of the shooting was the main topic of conversation during the May days following. Some of the Swedes who did not belong to the colony or who had left it, thought that Jansson had finally gotten what he deserved. Olaus Gus Petterson of Rock Island wrote to friends in Sweden that Jansson and the Jansonists were unpopular in the area: "Jansson was so disliked by Americans and Swedes that if someone admitted that he was from Bishop Hill, the Americans would not give him work." [10] Someone who signed himself "W" offered the readers of the *Gem of the Prairie* what seems to have been a view widely held: "That Root is guilty of deliberate homicide is beyond all question. According to Christian law, he is highly culpable. As to the feelings and principles of most men, he is partially if not wholly justified." [11]

Heated arguments about the case were held in the homes and meeting places throughout the county, some people defending Eric Jansson and the colonists, and some accusing Jansson of having stolen a man's wife away in order to have sexual relations with her. On May 25, W. W. Drummond, the attorney from Toulon who had befriended the colony in the past, wrote a letter to Governor French to acquaint him with the true state of affairs. He spoke with affection of "my deceased personal friend, Rev. E. Jansson," and said that sentiment seemed to be swinging against him. There was even a rumor that State's Attorney Reynolds would refuse to prosecute Root, but would instead bring the colony to trial. Drummond hastily added that he believed the rumor "to be false in substance and in fact," but he thought there was some truth to the rumor that "the Methodist Church is advocating Root's course as a Murderer." [12]

The atmosphere was too charged to find jurors who had not already made up their minds about the case, and though Judge Wilkinson issued a summons to bring John Root from his cell at Toulon to stand trial at Cambridge in October, 1851, the hearing had to be postponed again.[13] A change of venue was granted, and John Root was transferred to the prison at Rock Island, where he brooded another year before being brought to trial before the Knox County Circuit Court in the September term, 1852.

Even at Knoxville, and two years after the shot had echoed through the corridors of the Cambridge Courthouse, it was almost impossible to assemble an impartial jury. During Thursday, Friday, and Saturday, September 16, 17, and 18, nine panels were summoned and dismissed before twelve tried men and true could be seated.[14] The trial itself did not last long. Again Harmon G. Reynolds represented the People, and Julius Manning the defendant. Manning argued that the shooting was in no sense a deliberate murder, but was instead the natural response of a man whose wife's affections had been alienated by Eric Jansson. The jury agreed with him.[15] "We the Jury," reported the foreman, "do find the Defendant John Root Guilty of Man Slaughter and do fix his period of Imprisonment in the State Penitentiary at Two years."[16]

Root was taken to the prison at Alton, Illinois, and began his sentence by spending five days in solitary confinement. He was ordered to spend two years at hard labor and to pay all the costs of the trial. In May, 1853, while her husband was still in prison, Lotta Root sued for divorce and for the custody of their son, John. Judge Ira O. Wolkinson, who was then on the bench of the Tenth Circuit Court, sitting in Cambridge, heard the case. Lotta's grounds for the petition were simply the "infamous crime" her husband had committed, and Judge Wilkinson, ignoring the verdict of manslaughter which the court had previously found, had now no reason to disagree with her. He ruled that the bonds of matrimony which had united the two should now "be forever dissolved and for nothing held," and that the custody of John, who was two years old, should remain with his mother. He ruled also that John Root should pay the cost of the action, but Root pleaded (through Attorney Manning) that he was in jail, had no money, and had no prospect of raising any.[17]

Root's American friends immediately began agitating for his release. Benjamin D. Walsh was a cultivated and eccentric Englishman who owned a three-hundred-acre farm at Red Oak adjoining the colony lands, and who had developed a dislike of the Swedes because they lied in court, considered him one of the *otrogna* ("unbelievers") and dammed up the Edwards River, thus causing (so he thought) the region to become a breeding place for malaria. He had moved to Rock Island after the cholera epidemic in 1850. On January 2, 1854, he wrote a letter to Gov. Joel A. Matteson, reviewing what he took to be the facts of the case, and pleading that Root should be pardoned. "I believe the truth of the matter was," he wrote, "that when the woman was with Root she was willing to live with him; but that when the prophet got hold of her, she was persuaded that she would not go to Jesus Christ unless she abandoned her unbelieving husband." Walsh thought Jansson's motive in the affair was simply to avoid losing another hand for work in the fields. He pointed out that Root was a gentleman. "I have seen letters of his, covering four sides of a sheet of paper, which are perfectly faultless both in grammar and spelling." Root had the sympathy of his intelligent neighbors. "The almost universal feeling," Walsh wrote, "with educated men in this section of the county, who are acquainted with the circumstances of the case, is that if they had been in Root's place they would have acted as he acted." [18] At the same time, the governor received a petition for Root's release signed by over six hundred citizens of Rock Island.

The governor decided to agree. Though he had not had time to make the action official, he had decided to commute Root's sentence, but only on condition that Root would never again set foot on the soil of Illinois. Governor Matteson's daughter, Lydia, and a friend, Julia Rossiter, went to the prison in Alton to tell Root the good news. In a letter dated March 3, 1854, Lydia told her father about the visit. The prisoner she said, "looks wretchedly," in fact, he looked as though he would not live. When Julia told Root that he would be pardoned, but on condition that he leave the state, his chin quivered, his eyes filled with tears, and his body began to shake. "He told us that he had only seven months to remain in Penitentiary and scarcely expected to live that time—and he felt that he ought to be allowed to return to this state after his time had expired at any rate.—He seemed to think it hard to leave Illinois as he had many warm and ardent friends in this state in Rock Island, he would at least like to go there and die. He spoke very feelingly of his little son who he has not seen since he was four months old." [19] Lydia begged her father to remove the condition from the pardon. Four days later, on March 7, 1854, Governor Matteson signed a pardon for Root with no mention of banishment, and Root went off to Rock Island. Sometime later he moved back to Chicago.

Root's means of support during his years in Chicago after his release from prison have remained a mystery. Pastor Unonius has left us an account of him during this period.

> Having served his term in prison, he made his home in Chicago where he died a couple of years later, sometimes blaspheming God, sometimes in despair crying to him for mercy. So far as I know, he never troubled himself about his wife and child during that time, but left them to live undisturbed in Bishop Hill. The man was one of the most dangerous, most savage, and crafty persons I have ever known. Once, he was instituting a lawsuit against my friend Schneidau, he said quite frankly and deliberately, "Look out, Captain, I have committed one murder and won't hesitate to commit another." How he maintained himself after he was released from prison was a riddle to everybody. He was never known to work; but during the years he lived in Chicago many burglaries were committed in which he was suspected of being implicated, and not without cause.[20]

All the comments about him which have survived testify to the fact that he lived in dissipation and ill health. "He is said to have lived a wretched life after he left the penitentiary," said a local Swedish newspaper, *Hemlandet*, "curses flowing from his lips until the end, and his dead body was said to have been disgusting even to look at." [21] When the end was near, he asked to see a Methodist minister, Erik Shogren. Hair-raising stories about his deterioration were circulating in the Swedish colony, and Shogren went to call on him with some apprehension. This is his account.

> Ruth [Root] sent for me a few days before he died, and when I came, he asked me if I would accompany his body to the grave. I could not refuse. I visited him a few times and found him always very cold and unfeeling, but he never showed any other signs of fear, and never said a word about anything, whether of one thing or another, except that when I asked him why he wanted me to conduct his internment, he answered at once, in his own words, "I believe that you fear God. There are other clergy, but I do not like the way they live. You go with me to the grave" The day of the funeral came, and I think the whole of the Swedish community in

> town was in the funeral procession. At the grave I announced the funeral sermon would be preached in our church in the evening. Most of them came to the service. I preached on the text, "All flesh is grass." [22]

Lotta Root lived out the rest of her days pleasantly on her own farm, two miles from Bishop Hill, together with her son. In August, 1889, she was annoyed at persistent stories which she knew to be unfair to Eric Jansson. She drew up an official statement, reciting again the basic facts of the marriage and the murder, and emphasizing again that it was her decision to stay at Bishop Hill. "I say and declare," she said, "that my cousin, Eric Jansson, never behaved in an indecent or improper manner towards me, and the rumors about this are false." [23] She died in Galva in 1905 at the age of eighty. It was her joy to see her son, John, grow up to become a distinguished Henry County lawyer, who lived near her in Galva, Illinois, and who was highly esteemed both by the Americans and the colonists. He was the founder and the first president of the Old Settlers' Association.

Eric Jansson's widow, Sophia, lived for a while with her son at the Shaker colony in Pleasant Hill, Kentucky, but the boy proved to be wayward. Sophia ran a boarding house in Galva for a few years, but it did not succeed, and she died in the county poorhouse not far from Bishop Hill in 1888. She was buried near Eric Jansson's grave under a large cottonwood tree.[24]

In 1863, the prophet's son, Capt. Eric Johnson, as he was always called after the Civil War, married Mary O. Troil, daughter of the colony silversmith, and they had three sons and three daughters.[25] He became well known as a newspaper and magazine publisher, and when he died in 1919 at the age of eighty, John Root served as one of the pallbearers. When Baron Axel Adelswärd visited Bishop Hill in 1856, he found Eric Jansson's daughter, Mathilda, teaching in the local school, though she was only fourteen years old.[26] She married an officer in the Union army, Capt. A. G. Warner, while he was still in the service. After his discharge, he served for a while as sheriff of Henry County, and then he and his family moved to a farm in Iowa, where a son and a daughter were born.[27] After the captain died, Mathilda married Andrew Rutherford, and lived in Iowa until her death in 1926.

The colony recovered slowly after Eric Jansson's death, but a glory seems to have departed from the hill. A charter was granted by the Illinois legislature on January 17, 1853.[28] Under the guidance of Jonas Olsson, who had asserted his leadership and taken command together with seven trustees after his return from California, Bishop Hill prospered financially until the economic crisis of 1857. But the village became hopelessly divided between two parties, and in 1862 the communal idea was abandoned, each member receiving a fair share of the whole. Most of the colonists joined the local Methodist church, which was organized in 1864, but others joined the Second Adventist Church, which began holding services in the Colony Church in the late 1860s. A Swedish Mission church was formed in Bishop Hill in the 1880s, and continued until about 1900. After the passion of perfectionism had died down, the true believers seemed puzzled and apathetic. In the words of Edna St. Vincent Millay,

> What care we that the knowing ones will know
> We rose from rapture just an hour ago?

A Living Museum

> Because of its importance in our history, the Bishop Hill Memorial can be made a mecca for students, for our school children and the general public of the entire state. It can become another New Salem, on a smaller scale, a living museum of the life of its founders.
>
> —Governor Dwight Green, in the 1946 Centennial Address

The men who walked through the high grass of the Illinois prairies a century and a quarter ago are dead now. And the women who came with them, striding at their side or riding on wagons, are gone too, and so are the children they carried in their arms and spoke to in the accents of Hälsingland, and Dalarna, and Västmanland. But one lifetime ago some of them were still alive, sitting quietly in the warm sun, ready to talk about God and the odd things that happened when they set out to serve Him in a new land.

A strange feeling comes over one visiting Bishop Hill today, looking at the structures they built, sitting in the pews of their church, searching for traces of their brick kiln and boundary wall. One communes secretly with vanished people whose names have become familiar and whose very faults have become endearing. But even for those thousands of summer visitors who have never heard of Sophia Schön or Jonas Olsson, Bishop Hill has become an interesting historical site, holding in trust a bit of the American past, like a Civil War battlefield, or like a house where a president was born. In a country where all are immigrants, or are descended from immigrants, there is a universal interest in pioneers who came with some sort of dream and shaped it into a city which is vestigial now, but which once, like Camelot, was alive and full of the sounds of living. Bishop Hill can be used as an icon of the great folk emigrations and the founding of a new country by people who could not even speak its tongue. Unfortunately, the Janssonists asked more of the new land than any earthly land could ever give.

The story of the birth and death of Eric Jansson and the colony he founded is a tale worth telling, whatever its basic meaning or the lesson one can learn from it. But there must be some way of explaining why the story is memorable. What can we make of Eric Jansson and his followers apart from the physical traces they left and our sentimental memories? Did they make some kind of contribution to Swedish life? Did they have any influence upon Swedish laws or customs? Did they perhaps begin or continue the American myth, the dream of the great good land to the westward, from which Europe has never recovered? What was the impact of the Janssonists in America?

Such questions move us instantly from the world of objective history to the world of speculation, but they are worth answering for all that. Some charges against the Janssonists and some claims for them have no ground in fact. It has been said that because the Janssonists had been hounded out of Sweden by the application of the Conventicle Edict, the shameful quality of the act became visible to the Swedish people, who shortly afterward voted the edict out of existence. There were contemporaries who tried to use the expulsion of the Janssonists in this way. Pastor Anders Sandberg of Madesjö rose in the Parliament on December 23, 1847, to urge that the Conventicle Edict must go, since it had been used to drive out the Janssonists who were good men: "After studying what has been written regarding *läseriet* in northern Sweden, I have come to the conclusion that it was the existence of the Conventicle Edict of 1726 and its application which in no small measure brought forth the religious movement in Hälsingland, and that the provisions for punishment provided by the Edict harried several hundred Swedish citizens out of the country. They were not lazy men, nor vagrants, nor proletarians. They were self-supporting in their native land, and some of them were prosperous." Jonas Janzon, the associate minister in Linköping, agreed. "It would be better for us," he said, "if we had more of such *läsare* in our parishers." [1]

There were many in the Parliament who wanted to get rid of the edict, but showing pity for the Janssonists did not seem to them the way to achieve that end. What they had in mind was the

Swedish inscription on Eric Jansson's tombstone in Bishop Hill Cemetery. Courtesy of the Bishop Hill Heritage Association

substitution of the so-called "house-worship," which had been proposed as early as 1838, and which could be carried on with the blessing and help of the local clergy. The last thing they had in mind was the encouragement of heretical and Separatist enthusiasts. There were angry debates on the subject in Parliament on January 4, 8, and 12, 1848, during which the Janssonist offenses were rehearsed: the book burning, the violent resistance against Crown authorities, the bloody tumults, the breaking up of families. The respected Professor Johan H. Thomander took the floor and tried to repair the damage: "It is my opinion that Eric Jansson was a person of such character that many words should not be wasted on him before this august assembly. . . . What Eric Jansson has to do with the Conventicle Edict I do not understand. . . . Eric Jansson and his followers were heretical and it was proper that the 1726 law should be used against them." [2]

But it was too late. Having been reminded of the outrageous Janssonists, Parliament voted in favor of keeping the edict, partly, as Newman says, because of fear of the growing separatism in Norrland, but also, as Österlin adds, because of the fear of separatism in all of Sweden without adequate safeguards to protect the unity of the faith.[3] The *läsare* in Sweden who were not Janssonists were not helped by the sect. In fact, it is possible to argue that because of the Janssonists, they suffered under the Conventicle Edict for a decade longer. When the much hated legislation was wiped off the books, on October 26, 1858, the liberating power was not at all the Janssonists, but an unusual coalition of liberal and *läsare* forces.[4]

Some have said that Eric Jansson and his followers were responsible for the great folk movement from Sweden to America which went on during the rest of the century. Did he in some unintended way precipitate the great emigration? There is some reason to think so, and chronologically, at least, the thesis holds. Before the embarkation of the *Charlotte* in 1846, there had been only sporadic adventurers who had set out for the new land, though Peter Cassel led a small party to Iowa from Östergötland in 1845. After the Janssonists showed the way between 1846 and 1853, a stream of people followed them. Many letters came back from the colonists speaking in extravagant terms of the new Canaan, and the letters were passed between the red cottages of Västmanland, Dalarna and Hälsingland. "If any one thing spread the

America legend in Sweden," said Vilhem Moberg, "it must have been the letters from Bishop Hill in Illinois." [5] In addition to the twelve hundred Janssonists, there must have been many who did not share their religious faith, but who did share their sense of promise in the west and set sail to serve a more secular ambition. At least one party of Lutherans, led by L. P. Esbjörn, set out for America to win the Swedish apostates back to their mother church. And Esbjörn founded the Augustana church in America.[6]

There is a complicating factor with which this theory must deal, however. Though there were a great many letters sent back with reports of the Bishop Hill Colony, many of them were written by disillusioned settlers who wanted most of all to warn their Swedish friends not to make the same mistake that they had made.[7] Letters of this sort, passed from hand to hand and printed in such hostile papers as *Helsi* in Söderhamn and *Hudikswalls Veckoblad*, must have convinced many not to exchange their familiar and manageable problems for those of a strange and terrifying land. Conclusions from the fact that the tide of emigration followed close on the heels of the Janssonist departures may be a *post hoc ergo hoc* fallacy. Had the Janssonists never left, the emigration would probably have continued just as it did, and in even larger numbers. All that one can say with confidence is that the Janssonists did publicize the America dream, and may have something to do with its acceptance in Sweden.

What is clear is that the Janssonist movement was the lengthened shadow of an extraordinary man. He was what is called in Sweden an "eldsjäl," a burning brand, and he swept through the country like a fire raging through pine tops in a dry summer. He transfixed his listeners, many of whom were illiterate and culturally deprived, by speaking of a glory hidden from the rich and the learned, and he offered them the heady prospect of joining an elite corps entrusted with a cosmic mission. The fact that they had to preserve their perfect innocence with secrecy, deception, and flight served only to confirm them in their belief that the powers of hell were ranged against them. As long as their leader lived, they were impervious to such disasters as drowning and disease and death, knowing that these were to be expected as the stratagems of a devil who was outdoing himself because his foes were worthy. It was only after the echo of John Root's shots died out in the courtroom at Cambridge, and Eric Jansson lay motionless on the floor, that their confidence wavered.

When their leader died their sense of identity was threatened. The Janssonists were a paradigmatic sect, answering in every particular Ernest Troeltsch's classic definition of a sect. Unlike the church folk, the sectarians were a small group, aspiring after personal inward perfection, and aimed at direct personal fellowship between members of the group. They made no peace with the state, though they suffered violence rather than perpetrating it, and staked their hopes on a pure colony of separated believers. They thought the established church demonic, and hated its stylized buildings and set liturgies.[8] The whole machinery seemed to them to have been set up consciously to prevent the encounter of the suffering soul with its only hope, Jesus Christ, and the whole operation was managed by dubious professionals who seemed bent on silencing any amateurs gifted by the Holy Spirit. Eric Jansson was able to define his group in Sweden using his enmity against church and state to fix his own boundaries; but when the Janssonists came to America, where there was no church at all in the classical sense, and where the state was only a bemused spectator of these eccentric foreigners, his followers began to lose their sense of identity. It is probably inevitable that when the group at last disintegrated, it was not so much because of pressures from the outside, but because of unresolved hostilities within the group itself.

Eric Jansson was not only the chief officer in the sect. He was a copybook exemplar of what Max Weber called "the charismatic leader." Charisma is "a certain quality of an individual personality by virtue of which he is set apart from ordinary men and treated as though he were endowed with supernatural, superhuman, or at least specifically exceptional powers or qualities."[9] There can be no question of the hypnotic powers of this wheat flour salesman, who offered at first a business deal and then went on to convince twelve hundred customers to sell all their worldly goods, leave their native land, and follow him to America. As though they were following the charismatic model, they expected their leader to rise from the dead during the three days in which he lay in state after the courthouse shooting.

Sociological models might help a little to pluck out the heart of this mystery but psychological patterns are available as well. Eric

Jansson served his people in the role of what Sigmund Freud would call a tribal father. The simple folk who rallied around him needed such leadership to emerge from obscurity, and Eric Jansson offered himself as a father figure by means of which his family could identify itself. If Freud was right about such matters, there might even be unconscious erotic elements operative here, he being drawn to his followers and they to him in ways which both he and his rural Pietists would indignantly deny. Or perhaps it is helpful to think of him as what Carl Jung would call an "animus," the expressive leader who pointed to hidden depths of meaning, the master of fine words and ideas using "words," as Jung said, "full of meaning which purport to leave a great deal unsaid." The animus "must also belong to the 'misunderstood' class or be in some ways at odds with his environment, so the idea of self-sacrifice can insinuate itself. He must be a rather questionable hero, a man who had not really achieved much but who was filled with possibilities." [10] So defined, Eric Jansson was a classical animus—expressive, filled with vague hyperbole, in a frightening way both corrupt and corrupting, brilliant and unfulfilled, and hunted.

These are ways which are provisionally helpful in understanding Eric Jansson. But perhaps it is enough to think of him as a born leader, unacknowledged by the establishment because he had defied its rites of initiation and threatened its control, but yet a folk leader for all that, who ruled his people by the simplicity of his idea, by the strength of his convictions, and by the sheer audacity of his program. Like Freud's tragic leader he too suffered deeply and in the end seemed to take on himself the suffering of others, finally going to his death as a scapegoat (the Swedish "syndabock"). In a way which perhaps no one can understand who is outside the magic circle of mythic and cultic belief, there were those who knew that he would rise again, breaking the weak chains of death, just as there were those who knew that King Arthur would live forever in Avalon with only the very fairest of the elves. The Janssonists needed his return. By the sheer power of his magic he had transformed his followers from a nameless herd of farmers and sharecroppers into divine emissaries. But their position was precarious. Without his ringing words and flashing eyes, the program quickly dissolved into nonsense, and his people were soon arguing about who would have what share of the broken kingdom.

There are depths into which no one can peer, and Eric Jansson's power must remain something of an enigma to us. His faults were visible enough and often catalogued: an incredible self-confidence, which seemed to have been fanned into flame by the very need of his followers to be led; a theological naïveté, which caused him to grasp compelling ideas such as perfectionism, but without the qualifications which made these teachings creative; a blindness to the truth which was resident and available in other faith affirmations; a bigotry vividly demonstrated by the burning of the books. He managed to give his sect a fanatical unity, but at the same time deprived them of the corrective insights which those burning pages might have supplied.

Some of his great power came from his discovery of what Luther knew as well as he: that the devil was an active presence of evil in the world, prowling about and seeking whom he might devour. Jansson never recovered from the writhing demonic shapes which he had seen painted on the walls of the narthex on the church in Österunda. But he appropriated the idea in a form which Luther would have rejected. Luther saw that the demonic force was in all men, including himself, and the devil plied his trade in the souls of every man, tempting, deceiving, slaying, seducing. Therefore all men were subject to what Luther called the strange work of God—his wrath, the only form of his love which appears to fallen men. Jansson, on the other hand, thought that the devil was only a problem for other men, since he and his band of followers had achieved purity. It followed from his corrupted version of Lutheran demonry that he and his followers must seek a solitary place, where they could be uncontaminated by evil, and where they could celebrate together the ceremony of innocence. They discovered with a hurt surprise that the devil went to America with them and flourished even within the earth walls of Bishop Hill.

Eric Jansson shared with the Methodists their vision of a perfect holiness which God sometimes offered as part of the divine largesse to his children. There were curious affinities between Janssonism and other patterns of thought. He was in the tradition of Catholic mysticism, which affirmed the possibility of a divine presence unmediated by earthly forms, and uncorrupted by historical limitation. And he also was in the tradition of the romantic utopians who founded a series of colonies in America in the 1840s. What this wheat flour visionary saw was something missed by his more pragmatic neighbors: the possibility of time being invaded by eternity. He did not know and had no way of finding out except by suffering that there is indeed a point of intersection where time and infinity meet, far more rich and satisfying than moments pragmatists know, but not for all that the same as infinity itself. What could be known at that hallowed center was more like a hint and a promise, an earnest of blessings yet to come and not a fixed splendor which could be claimed and held as a simple possession. The truth he had to learn was that any historical situation was a mixture of possibility and limitation. So much was this Bible-hungry farm boy filled with a wild longing for holiness, that he did not notice the corruption even in his dreams. And when it became more and more clear that the dream was lost he retreated to the simpler certainties of his days as a salesman, at the end trying to move mats in St. Louis as he had once tried to move wheat flour in Hälsingland.

The fairest judgment of him is probably that offered by Fredrika Bremer just after his murder in 1850: "The man seems to have been an enigmatical character, half a deceiver and half deceived either by himself or by his demon." [11] There was in his whole movement a dangerous combination of strength and weakness, of surrounding circumstance and unfortunate decision. When the Mission Friend leader Peter Waldenström visited Bishop Hill in 1889, he quoted with approval a remark made by Wilhelmina Westerberg, at that time an old lady and mellowed since the days when she fought with Arenander: "In those days the people back home didn't know any better, and we didn't know how to tell them what we meant." [12] The guilt must be parcelled out, though surely not in equal portions. It is possible to read iconic significance in the remark made by an old lady to the wife of Archbishop Söderblom, who was visiting Bishop Hill in 1923. The old lady had come to the colony as a young girl and had now grown quite deaf. "Stand close to me," she said to Mrs. Söderblom, "and talk a little roughly. Then I might hear you." [13]

There was no doubt an historical necessity in Jansson's followers abandoning the utopian pretensions after his death, and returning as many did to the Lutheranism of their childhood. They knew by bitter experience what the pastor meant when he uttered the *iustis et peccator simul* formula from the Augustana pulpits of Andover and Galesburg. One could also perhaps have predicted that when the utopianism faded many Janssonists would turn with relief to the local Methodist church, where the familiar perfectionism was taught, but this time interpreted not as a simple possession but as aspiration and love. And no one need be in the least surprised that some of them, led by Jonas Olsson, also turned to the Second Day Adventists, clinging to the possibility of a perfect holiness, but not expecting it until some future time when the Son of Man would return to the earth in glory and claim his own.

No one who turned to another confessional body ever forgot Eric Jansson and his burning eyes. The Bishop Hill Colony was founded and flourished in the early years because of the hopes of this remarkable man, who had been granted a vision that human life stands under an ideal possibility purer than the classical reformers had ever thought possible. The colony floundered partly because of the inevitable utopian blunder: the inability to recognize the pervasiveness of evil and the provisional character of all human achievements. But it failed as well because of a simple theological error for which Eric Jansson must bear the major blame: the failure to see that perfectionism is creative as longing, but only as longing.

However, we do not today remember Eric Jansson because of his failures—his baffled plans and ruined hopes—nor do we think of him pitching to the floor of the courthouse in Cambridge. In a curious way the vision defeated is the thing that lives, and not the menacing forces that seemed in the end to destroy it. What clings to our minds is the fragile and at the same time tenacious dream of finding somewhere on earth a great good place, secure against ambiguity, secure even against the dark circling forces so evident in our time. The *Patria* he sought for is also our own. Let him only sneer who sought for it as bravely as did Eric Jansson.

NOTES

·

SELECTED BIBLIOGRAPHY

·

INDEX

Notes

1. THE VISITATION

1. Jansson was writing the second part of his "Autobiography" on May 2, 1850, just a few days before he was murdered, and he may have forgotten the exact date of his birth (pt. 2, p. 25). The "Dop Bok för Biskopskull Församling," in the Uppsala Landsarkiv, is not likely to be wrong about this, and it says that Eric Jansson was born on December 21, 1808, and was baptized on December 22 (p. 30).

2. From a manuscript entitled "Rörande Routh's Hustru" [Concerning Root's Wife] in the possession of Emmelyne Arnquist Hedstrom of Galva, Illinois, p. 18. The MS will hereafter be cited as "Hedstrom MS." Anna Maria Strâle was born in Torstuna on October 11, 1806, the daughter of a soldier, Peter Sten Strâle, and had served one year as a servant in Österunda before moving north to Hälsingland. After becoming a convert to Janssonism, she moved back to Österunda and lived in the home of Eric Jansson's parents, Klockaregården (*Erik-Jansonisternas historia* [Galva, Ill., 1902], p. 13.) See Nils William Olsson, *Swedish Passenger Arrivals in New York, 1820–1850* (Chicago, Ill., 1967), p. 93.

3. "Autobiography," pt. 1, p. 9.

4. Preface to bk. 1, p. 3.

5. "Autobiography," pt. 1, p. 9.

6. Gustaf Wingren, *Luther on Vocation*, trans. Carl C. Rasmussen (Philadelphia, 1957), pp. 3, 64–65, 73.

7. Maja Stina was the daughter of Lars Jansson, a farmer from Rung, Torstuna Parish, and of Kajsa Larsdotter. She married Eric Jansson in Torstuna on November 1, 1835. Her father came to America with a party of Janssonists on September 15, 1846 (Olsson, *Swedish Passenger Arrivals*, pp. 96–97).

8. "Autobiography," pt. 1, p. 11. Col. 2 goes a long way toward the antinomianism which Eric Jansson did not quite dare to spell out: "And you, being dead in your sins and the uncircumcision of your flesh, hath he quickened together with him, having forgiven you all trespasses. Blotting out the handwriting of ordinances that was against us" (vv. 13–14). Jansson's claim was not of course that he and his followers could sin all they pleased, but rather that nothing they did was sinful.

9. "Torstuna Dödsboken, 1836–37," Uppsala Landsarkiv.

10. The associate minister Jacob Risberg was born in Gävle in 1809, and so was one year younger than Eric Jansson. He was ordained in 1833 (*Upsala ärkestiftets matrikel, 1850*, ed. John Eric Fant [Uppsala, 1850], p. 177).

11. "Autobiography," pt. 1, p. 11. Emil Herlenius, who wrote one of the primary books about the Janssonists, said that Jansson was especially offended by a doctrine which was popularly held at Torstuna during this time, and which held that souls lived on after the body's decay and sometimes returned to earth. There were many readers of Christian Bastholm, *Philosophiska bref angående själens tillstånd efter kroppens död*, trans. Carl Peter Blomberg (Stockholm, 1794). See Emil Herlenius, *Erik-Jansismens historia: Ett bidrag till kännedomen om det Svensk sektväsendet* (Jönköping, 1900), p. 9, n. 1.

12. Estenberg said that he had never seen such ecstasy in a prayer (*Erik-Jansismens historia*, p. 16). This rare and important book, in which the colonists themselves tell their own stories, was published with no hint as to the editor, place or time of publication. George Stephenson in his biography identifies the editor as "[Westerberg, fru]," and says it was published in Gavla, Illinois, in 1902. The date and place are acceptable guesses, but he knows of no evidence which points to "Fru Westerberg." He probably meant Wilhelmina Ohrström, who married the wagonmaker John H. Westerberg in 1851. But on p. 56 the book refers to her as being "still alive." If she edited the book, the reference is puzzling, and she must have been very old when the book was published.

13. "Autobiography," pt. 1, p. 14.

14. *Erik-Jansonisternas historia*, pp. 56–57.

15. Ibid. Anders Andersson, Torstuna Parish, Västmanland, was an ardent Janssonist. He and his family sailed for America on the *Patria*, arriving in New York on August 21, 1846 (Olsson, *Swedish Passenger Arrivals*, pp. 80–81).

16. Manuscript letter in the Uppsala University Library, dated June 30, 1845.

17. "Österunda uttflytrningsbok," in the Uppsala Landsarkiv.

18. The question of who proposed to Jansson that he visit Hälsingland is what the French call *discutable*. Twice in his "Autobiography" Jansson said that Risberg advised him to go there, and P. A. Huldberg, editor of *Hudikswalls Veckoblad*, blamed Risberg (who was then pastor at Hög, near Hudikswall) for the whole uprising. But Risberg indignantly denied the story. He said that so far from agreeing with Jansson's views, he was opposed to them, and "openly worked against them." He said that he had warned Jansson that the proposed trip would be good neither for Jansson nor for Hälsingland (*Hudikswalls Veckoblad*, February 1, 1845). At his trial held in Forsa in October, 1845, Jansson told the court that Anders Arquist, associate minister in Enånger, had been the one who persuaded him to come to Hälsingland (ibid., October 11, 1845). Editor Huldberg was no doubt close to the truth when he wrote that Pastor Risberg had written a glowing letter of recommendation to Pastor Arquist of Enånger (ibid., February 8, 1845). Wilhelmina Westerberg must be given the final word: "Pastor's assistant Ahlquist [surely she means Arquist] who served in the Enånger Parish, and with whom Eric Jansson was acquainted, invited him to come for a visit" (*Eric-Jansonisternas historia*, p. 115).

2. THE SALE OF WHEAT FLOUR

1. See the charming account of farm life in Hälsingland at this period in *The Early Life of Eric Norelius (1833–1862)*, trans. Emeroy Johnson (Rock Island, Ill., 1934), pp. 81 f.

2. When the Mission Friend revivalist E. A. Skogsbergh visited Bishop Hill in 1878, several of the original settlers were struck by his physical resemblance to Eric Jansson (E. A. Skogsbergh, *Minnen och uplevelser* (Minneapolis, Minn., 1925), p. 186).

3. Bror Alstermark, *De religiöst-svärmiska rörelserna i Norrland, 1750–1800: P. I. Herjeådalen och Helsingland* (Strängnäs, 1898), p. 109; and A. G. Sefström, *Några blad till historien om läsarne, med afseende fastadt på de inom Helsingland vistande* (Falun, 1841), p. 52.

4. Jonas Olsson was born in Söderala Parish, Gävleborg *lan*, in 1802 (Olsson, *Swedish Passenger Arrivals*, p. 87).

5. Nathanel asked Philip, "Can there any good thing come out of Nazareth?" (John 1:46).

6. *Upsala ärketsiftets matrikel, 1840*, ed. J. E. Forsell (Uppsala, 1840), pp. 144–45.

7. Henrik Gladh, "Till hälsingeläseriets och Erik-Jansismens karakteristik," *Kyrkohistorisk årsskrift* 47 (1947): 186–212. Gladh printed only the relevant parts of this important document, the whole of which is in the Uppsala Landsarkiv. Replies came back from some forty parishes, and Baron Lagerheim sent them on to the archbishop. A summary of the material also went to the minister for Ecclesiastical Affairs. Twenty parishes reported that they had no Janssonists at all in their neighborhood. Adult Janssonists said to be still in Sweden numbered 649.

8. C. G. Blombergsson became the Janssonist printer for one hectic year, 1846, using his press in his home, not far from that of Jonas Olsson, at Ina (see Olsson, *Swedish Passenger Arrivals*, p. 71).

9. In his earliest publication, Eric Jansson had advised his readers to speak sternly to anyone not following rigid devotional practices: "If you meet anyone who is a father or mother who is not helping the household to pray," he said, "you ought to tell them what you think of them at once, without worrying about personalities" (*Några ord till Guds församling* [Söderhamn, 1846], p. 4).

10. Born in 1801, he served as associate pastor in Norrala and Trönö, 1838 to 1853 (*Upsala* ärkestiftets herdaminne, 1843, ed. John E. Fant and August T. Låstbom [Uppsala, 1843], 2:304; *Upsala ärkestifets herdaminne 1893*, ed. Ludvig Nyström [Uppsala, 1893], 4:316).

11. Herlenius, *Erik-Jansismens historia*, p. 15, n. 2.

12. E. J. Ekman, *Den inre missionens historia* (Stockholm, 1898), 1:796.

13. J. O. Nordendahl was born in 1805, and ordained in 1840. He had been pastor at Enånger and Njutånger since 1840. Jansson's friend Anders Henrik Arquist was born in Alra in 1811 and was ordained in 1834 (*Upsala ärkestiftets matrikel, 1850*, pp. 110, 354).

14. He was born in 1796, and died in 1849 (ibid., p. 357, and *1893*, p. 133 ff.). A letter from him to George Scott is published in Gunnar Westin, *George Scott och hans versamhet i Sverige* (Stockholm, 1929), 2: 382–84.

15. Clas Erik Claesson (1803–71) was the Mr. Chips of Hudikswall from 1832

to 1870. A bachelor deeply devoted to his pupils and very devout, he was popular in the community, and when he died the townsfolk erected a monument over his grave in St. Jacob's churchyard (*Upsala ärketiftets herdaminne, 1893*, 4:219).

16. Carl Ludvig Boström had been pastor in Hudiksvall and Idenor since 1839 (ibid., *1843*, 2:219). A note written beside his name in this book in the Uppsala Landsarkiv says, "He hanged himself in 1851." When the Ministry of Justice closed down Blombergsson's printing shop in 1846, Boström was given the unpleasant task of enforcing the order (*Hudikswalls Veckoblad*, February 7, 1846).

17. Herlenius, *Erik-Jansismens historia*, p. 16.

18. Ibid., p. 16.

19. "Autobiography," pt. 1, p. 37.

20. Johan Eric Fillman, born in Hille, 1800, had come to Hamrånge and Axman's Bruk in 1831 as a schoolmaster and associate minister (*Upsala ärkestiftets herdaminne, 1843*, 2:28).

21. She was born Katrina Wäxell in Tyby, Mo Parish, in 1809, the daughter of the church sexton (Olsson, *Swedish Passenger Arrivals*, p. 87).

22. P. N. Lundqvist, *Erik-Jansismen i Helsingland. Historisk och dogmatisk framställning jemte wederläggning af läran* (Gävle, 1845), p. 93.

23. Born in 1796, he suffered from poor health all his life and died in 1849 (*Upsala ärketiftets herdaminne, 1893*, 4:133–34).

24. A letter which he wrote to George Scott on July 1, 1842, is printed in Westin, *George Scott*, 2:368–69.

25. Anders Gustaf Sefström was born in 1790, and became minister in Rogsta and Ilsbo in 1832, and in Norrbo and Bjuråker in 1841 (*Svenska män och kvinnor* [Stockholm, 1942–45], I: viii). See the account of his ministry in Emil Petterson, "En Ljusdalpräst," *Julhälsningar till församlingarna i ärkestiftet, 1929* (Uppsala, 1929), pp. 27–35. His preaching brought on the crisis which led to the conversion of Oscar Ahnfelt, Rosenius's musical helper in Stockholm (Karl A. Olsson, *By One Spirit* [Chicago, 1962], p. 56).

26. Peter Käck, a soldier in the Hälsinge regiment, was born in 1802. He sailed with other Janssonists from Gävle on the *Solide* late in 1846 (Olsson, *Swedish Passenger Arrivals*, p. 129).

27. Herlenius, *Erik-Jansismens historia*, p. 18.

28. Letter written March 11, 1842, to George Scott (Westin, *George Scott*, 2:298). Jansson repeated the charge in *Ett ord i sinom tid*, p. 4.

29. *Ibid.*, II, 242. Jansson's gossip about Sefström is in the "Autobiography," pt. 2, p. 17.

30. Olsson, *Swedish Passenger Arrivals*, p. 107. During the voyage across the Atlantic, Fru Hebbe and Sophia Schön insisted that he be deposed from the ministry, but when they arrived in New York he took up his duties again (Albin Widén, *När Svensk-America grundades* [Borås, 1961], pp. 33–34, n. 7).

31. Olof Jonsson from Valla, Söderala Parish, advertised an auction for the sale of his possessions before leaving for America on June 2 and 5, 1846 (*Kuriren* 25 [1871]). He became a trustee of the Bishop Hill Colony (Olsson, *Swedish Passenger Arrivals*, p. 109). Another man from Valla, Jacob Jacobsson, also became a trustee (Eric Johnson and C. F. Peterson, *Svenskarne i Illinois* [Chicago, 1880], p. 313). Nils Hedin was a tailor from Hede Parish, Jämtland (Olsson, *Swedish Passenger Arrivals*, pp. 104–5).

32. Anders Johansson (1775–1867). Archbishop Wingård spoke of him as a "Johannestyp" (*Upsala ärkestiftets herdaminne, 1842*, 1:473; *1893*, 4:179).

33. Letter dated April 10, 1843, in the "1843 Expeditionsbok" in the Uppsala Landsarkiv. Risberg returned to Hälsingland, but he had an unhappy career. He spent twenty-seven years as an associate minister, was so poorly paid that his home at Åkerby was heavily mortgaged, and when he died in 1863 he left his wife and child deeply in debt (*Upsala ärkestiftets herdaminne, 1893*, 4:608).

34. *Hudikswalls Veckoblad*, January 18, 1845, and February 1, 1845.

3. THE WORLD, THE FLESH, AND THE DEVIL

1. Set Rimbe, "Frälsaren på Stenbo," *Julhälsning till Forsa församling* (1954), p. 10.

2. But in 1848 the Grip family had become Methodists and were living in Galva, Illinois (Olsson, *Swedish Passenger Arrivals*, pp. 118–19).

3. Lundqvist, *Erik-Jansismen i Helsingland*, p. 16. See also O. A. Wallin, *Tidsbilder från Järvsö* (Stockholm, 1903), p. 72.

4. *Invandrarna* (Stockholm, 1952), pp. 28–31.

5. Herlenius, *Erik-Jansismens historia*, p. 15.

6. Lundqvist, *Erik-Jansismen i Helsingland*, p. 31.

7. *Upsala ärkestiftets herdaminne, 1842*, 1:286–97; *1893*, 4:104–8. See Torsten Bohlin, "Lars Landgren i Delsbo: Ett 100-årsminne," *Julhäsning till församlingarna i ärkestiftet, 1944* (Uppsala, 1944), pp. 48–58.

8. Bos-Karin was the daughter of Erik Olofsson and Margta Ersdotter. Her father was a cripple. She had three brothers (*Sveriges släktregister: Bjuråker, Delsbo och Norrbo socknar* [Stockholm, 1949], p. 427).

9. Ingvar Jonsson, "Fäbodbebyggelsen i Hälsingland," *Hälsingerunor: En hembygdsbok, 1964–65* (Malung, 1964), pp. 157–66.

10. This must be the Pehr Fors to whom greetings were sent by Eric Pehrsson from Center Township, Iowa, in a letter dated December 23, 1869 (Helmer Johansson private collection).

11. Kjerstin Andersdotter, a devoted Janssonist, was married to Erik Erickson-Boström, a bricklayer from Källeräng, Delsbo Parish. He was one of Bos-Karin's brothers (Olsson, *Swedish Passenger Arrivals*, p. 149).

12. Lundqvist, *Erik-Jansismen i Helsingland*, pp. 27–29. The six were Marta Jonsdatter Hjelm, Brita Pehrsdotter, Sigrid Ersdotter, Sigrid Olsdotter, Anna Pehrsdotter, and Isaac Rudolphi. Pastor Landgren testified that the witnesses called were all qualified, since they were members of his parish, had been confirmed, and were in good standing (ibid.). He quoted Swedish law for refusing to take testimony from any of the Janssonists, citing *Rättegång* B. 17, chap. 7: that no one could testify who had a part in the matter under trial, or could be benefitted or harmed by it (*Hudikswalls Veckoblad*, November 8, 1845).

13. *Hudikswalls Veckoblad*, August 25, 1846.

14. "I båda dett ena och dett andra."

15. "Autobiography," pt. 2, p. 29.

16. Lundqvist, *Erik-Jansismen i Helsingland*, p. 165.

17. Herlenius, *Erik-Jansismens historia*, p. 27.

18. *Ett ord i sinom tid*, p. 3.

19. Conversations with Philip Stoneberg, Stoneberg Collection, Knox College Library, Galesburg, Illinois. Bos-Karin died on May 6, 1887, at the age of seventy-one (*Sveriges släktregister*, p. 427).

20. Lundqvist, *Erik-Jansismen i Helsingland*, p. 35, n.

21. Herlenius, *Erik-Jansismens historia*, p. 21.

22. Olsson, *Swedish Passenger Arrivals*, pp. 104–5.

23. Ibid., p. 109.

24. Per Adolf Huldberg was born in Uppsala in 1815 of German ancestry. In 1836 he moved to Falun and founded the newspaper *Fahlu Tidning*. A youthful disciple of George Scott, he joined Gustava Rohl in the temperance struggle at Falun, and became more and more liberal in his politics (Emil Herlenius, *Religiösa rörelser i Falun under mediet av 1800-talet* [Falun, 1842], pp. 3–4, 10).

25. We do not know which book Jansson refers to, but it probably was *Några ord till Guds församling*, in which he says, "I shall stop writing now, and let the reader be satisfied with this first venture into print" (p. 28). The book was published by C. G. Blombergsson, the Janssonist publisher in Söderala, in 1846.

26. Holograph letter in the manuscript collection, Royal Library, Stockholm.

27. "Autobiography," pt. 2, p. 33.

28. Holograph letter in the manuscript collection, Royal Library, Stockholm.

29. See Nils C. Humble, *Två hälsingesocknar* (Gävle, 1934), pp. 228–30.

4. LIFE WITHOUT SIN

1. *Förklaring över den heliga skrift, eller katekes affattad i frågor och svar* (Galva, Ill., 1903). This is a reprint of the first edition published in Söderhamn in 1846. Hereafter it will be cited as *Catechism*.

2. C. A. Cornelius, *Svenska kyrkans historia efter reformationen* (Uppsala, 1887), p. 205.

3. From the Forsa *Dombok*, cited by Herlenius, *Erik-Jansismens historia*, p. 127.

4. *Treatise on Grace*, qu. 109, art. 2.

5. *Några sånger, samt böner, Författade af Erik Janson* (Galva, Ill., 1857), p. 11. This is the second edition, reprinted with the addition of some psalms from the first edition of 1846.

6. *Catechism*, p. 79.

7. Ibid., p. 80.

8. *Domboken, Thorstuna, 1845*, sig. 63. She later added an explanatory note: "I hold Erik Jansson not less than God: 'as the Father has sent me, so I have sent you' " (sigg. 66b and 67).

9. Favorite Janssonist proof texts were Heb. 13:20; 1 Thess. 5:8; Matt. 22:37–39. John was the favorite book. 1 John 3:6–10 is paraphrased in *Några sånger*, pp. 79–80.

10. An unpublished letter dated December 18, 1843, in the Royal Library, Stockholm.

11. John Edward Christopher Hill, *The World Turned Upside Down: Radical Ideas during the English Revolution* (New York, 1972), pp. 133–34. Norman Cohn, *The Pursuit of the Millennium* (London, 1970), p. 287.

12. MS letter from a colonist, E. Lindstrom, to J. E. Ekblom, June 3, 1867: "There is absolutely no difference in beliefs or formal confession between us and

the Methodist Episcopal Church in this country" (Fröken Lindewall Collection, Uppsala Landsarkiv).

13. *The Works of the Rev. John Wesley, A.M.*, ed Thomas Jackson, 3d ed. (London, 1829), 8:238. Wesley's fullest account of his perfectionist doctrine is in *Works*, vol. 11, *A Plain Account of Christian Perfection*, pp. 366–446; and in vol. 6, *On Perfection*, pp. 411–24. See also R. Newton Flew, *The Idea of Perfection in Christian Theology* (London, 1934), pp. 314 ff. I am indebted in my account of Wesleyan perfectionism to Philip Watson, sometime professor of theology at Garrett-Evangelical Theological Seminary, Evanston, Illinois.

14. *Works*, vol. 20, *Treatise on Baptism* (1756), pp. 188–201.

15. *Works*, 8:364 and 11:394, 396, 419.

16. Ibid., 11:417.

17. Ibid., p. 402.

18. See Westin, *George Scott*, 1:231–45.

19. Scott's manuscript "Journal of Sermons, Etc." has a note dated November 2, 1831: "Swedish preaching commenced" (Uppsala University Library).

20. Michael A. Mikkelsen, *The Bishop Hill Colony: A Religious Communistic Settlement in Henry County, Illinois*, Johns Hopkins University Studies in Historical and Political Science, ser. 10, no. 1 (1892; reprint ed., Philadelphia, 1972), p. 20.

21. George Scott, *Några ord om Wesleyanska Methodisterna och deras lärasater* (Stockholm, 1833), p. 9.

22. Lecture notes, Uppsala University Library.

5. RIGHTEOUS AND SINFUL AT ONCE

1. "Religionsvärmeriet i Norrland," *Aftonbladet*, February 8, 1845. The late Johan Ohlsson, Alfta's local historian, said that 480 people left Alfta with the Jansonists (*Alfta förr och nu* [Ljusdal, 1961], p. 23).

2. February 8, 1845.

3. *Hudikswalls Veckblad*, February 27, 1845.

4. *Evangelical Review* 1 (April 1850):572.

5. P. A. Åkerlund, *Om kyrkans och statens enhet* (Stockholm, 1854), p. 80.

6. Quoted by Henrik Gladh, *Lars Vilhelm Henschen och religions-frihetsfrågan till 1853* (Uppsala, 1953), p. 162, n. 1.

7. Åkerlund, *Om kyrkans och statens enhet*, p. 84.

8. Alstermark, *De religiöst-svärmiska rörelserna i Norrland*, p. 124.

9. Quoted by Gladh, *Lars Vilhelm Henschen*, p. 162.

10. Gladh pointed out this shift (ibid., p. 159).

11. See Anna Söderblom, "Läsare och Amerikafarare på 1840-talet; brev, protokoll m.m. om Erik Jansismen," *Julhelg för Svenska hem* (Stockholm, 1925), pp. 80–93.

12. *Grundlagen till läran om kyrkan* (Gävle, 1856), cited by E. Newman, *Gemenskaps och frihetssträvanden i Svensk fromhet liv, 1809–1855* (Stockholm, 1939), p. 39, n. 2.

13. Gladh, "Till hälsingläseriets och Erik-Jansismens karakteristik," p. 194.

14. The subtitle announces the polemical intention: *Historisk och dogmatisk framställning jemte wederläggning af läran* (Gävle, 1845). Lundqvist had the interesting notion that Janssonism was very much like Swedenborgianism (see his *Swedenborg*

och Bibeln [Söderhamn, 1850], p. 3). But Bailiff Ekblom of Torstuna thought the notion was nonsense ("Något om P. N. Lundqvists skrift Swedenborg och Bibeln," MS in the Fröken Anna Lindewall Collection, Uppsala Landsarkiv).

15. Gladh, *Lars Vilhelm Henschen*, pp. 164–65.

16. "Superbus primo excusator sui ac defensor justificatur" (Martin Luther, *Kritische gesammatausgabe* [Weimar, 1883], 3:288).

17. "Nescimus, quid Deus, quid justitia, denique quid ipsum peccatum est" (ibid., p. 106).

18. Vilhelm Dilthey, *Gesammelte Schriften: Weltanschauung und Analyse des Menschen Seit Renaissance und Reformation* (Göttingen, 1964), 2:162–202. Cf. Adolf Harnack's remark: "Through having the resolute wish to go back to *religion* and to it alone the Lutheran Church neglected far too much the moral realm, the Be ye holy, for I am holy" (*History of Dogma*, trans. Neil Buchanan (Boston, 1900), 7:266–67).

19. *The Social Teaching of the Christian Churches*, trans. Olive Wyon (New York, 1931), 2:498.

20. *Luther on Vocation*, trans. Carl C. Rasmussen (Philadelphia, 1957), pp. 55–56.

21. *The Cost of Discipleship*, trans. R. H. Fuller, London, 1948, p. 46.

22. *Social Teaching*, 2:493.

23. "Konventikelplakatets upphävande, ett gränsår i Svensk religions frihetslagstiftning?" (*Kyrkohistorisk årsskrift* 57 [1957]:136).

24. *A Christian Retrospect* (New York, 1851), p. 211.

25. *Hvilken är Sveriges religion?* (Stockholm, 1827), p. 25.

26. MS, "Six Lectures: Religion in Sweden," Uppsala University Library.

6. THE BURNING OF THE BOOKS

1. Set Rimbe, "Frälsaren på Stenbo," p. 10.

2. Anna Söderblom, "Läsare och Amerikafarare," p. 10. Olof Stenberg called himself "Stoneberg" in America. He became a trustee of the colony (Olsson, *Swedish Passenger Arrivals*, p. 245).

3. Lundqvist, *Erik-Jansismen i Helsingland*, p. 24.

4. Ibid., p. 25, n.

5. Ibid.

6. Flew, *Idea of Christian Perfection*, p. 339.

7. *Några sånger*, no. 1, v. 4, pp. 1–2.

8. Ibid., no. 4, v. 6, p. 8.

9. Lundqvist, *Erik-Jansismen i Helsingland*, p. 36.

10. This is grossly unfair, as Jansson must have known. In his sermon on the Eighteenth Sunday after Trinity, Luther said this: "In former times, all men, especially we monks, tormented ourselves with lengthy repetitions in reading and singing; yet our prayers were but chattering, as the noise of geese over their food" (Luther Kritische, 9:196). Luther's point is that *monks* chatter.

11. *Nordiska Kyrkotidning* (1845), quoted by Herlenius, *Erik-Jansismens historia*, pp. 28–29.

12. Lundqvist, *Erik-Jansismen i Helsingland*, p. 22. Pastor Landgren asked Jansson how he could criticize Luther's theology, and yet be dependent upon the Bible

which Luther had translated. Jansson told his followers that Landgren now wanted to take their Bible away. "You poor insect," shouted the exasperated Landgren, "you can't understand the Bible!" (Torsten Bohlin, *Lars Landgren, människan, folkuppfostraren, kyrkomannen* [Stockholm, 1942], pp. 97–98).

13. Lundqvist, *Erik-Jansismen i Helsingland*, p. 89.

14. Newman, *Gemenskaps och frihetssträvanden*, p. 84, n. 83.

15. *Sanna Christendom*, p. 3.

16. Ibid., p. 93.

17. Gabriel Rosen's introduction to Anders Nohrborg, *Den fallna människans salighets ordning* (Stockholm, 1771), p. iv.

18. Lundqvist, *Erik-Jansismen i Helsingland*, pp. 22–23.

19. Ibid., p. 61.

20. *Gefleborgs Läns Tidning*, June 19, 1844.

21. Herlenius, *Erik-Jansismens historia*, p. 29.

22. Lundqvist, *Erik-Jansismen i Helsingland*, p. 37.

23. *Gefleborgs Läns Tidning*, June 19, 1844.

24. *Stora Kopparbergs Läns Tidning*, June 27, 1844.

25. *Norrlands-Posten*, June 21, 1844.

26. Johan Ohlsson, "Om Erik Jansismen," pp. 81–95.

27. Lundqvist, *Erik-Jansismen i Helsingland*, p. 71, n.

28. Herlenius, *Erik-Jansismens historia*, p. 30.

29. *Hudikswalls Veckoblad*, April 5, 1845. The conservatives hated to see the old king die. When Archbishop Wingård met Henril Reuterdahl on the street he said "as long as the old lion lived, we could fight the pack. Now we must defend royalty even against the king" (*Arkebiskop Henrik Reuterdahls memoarer* [Lund, 1920], p. 212).

30. Reuterdahl called Silfverstolpe "a righteous and skilful man, but absolutely a stranger to church affairs" (*Arkebiskop*, p. 203). When he lost his office in 1848 he became governor of Västmanland and is remembered chiefly for his educational reforms.

31. *Stora Kopparbergs Läns Tidning*, November 28, 1844. This curt dismissal was not the exit the Janssonists heard about. They were told when the audience was over, the door lock jammed, and the king led Jansson out a back exit, after laughingly remarking that they were locked up together (Hedstrom MS, fol. 42).

32. Lundqvist, *Erik-Jansismen i Helsingland*, p. 40. Both followers and foes believed that the Janssonists were under the protection of the king (*Norrlands-Posten*, September 28, 1844).

33. Herlenius, *Erik-Jansismens historia*, p. 33.

34. *Aftonbladet*, October 25, 1844.

35. Sivert Erdahl, "Eric Janson and the Bishop Hill Colony," *Journal of the Illinois State Historical Society* 18 (October 1925):528. C. J. L. Almquist was an advocate of free love, and Nils Ignell was an ardent disciple of Schleiermacher.

36. *Stora Kopparbergs Läns Tidning*, November 14, 1844.

37. C. W. Berg and Amy Moberg, *Teckning af Carl Olof Rosenii lif och verksamhet, hos wänner tillegnad* (Stockholm, 1868), pp. 122–23.

38. Cited by Karl Linge, *Carl Olof Rosenius* (Uppsala, 1956), pp. 99–100.

39. Conversation between Jonas Olsson and Philip Stoneberg, Stoneberg Collection, Knox College.

40. Berg and Moberg, *Teckning af Carl Olof Rosenii*, p. 121.
41. *Pietisten* 14 (1878):173.
42. Lars Vilhelm Henschen to Augusta Henschen, November 29, 1844. Cited in Gladh, *Lars Vilhelm Henschen*, p. 104.
43. *Korsbloman* (1922), p. 127. Cited by Gladh, *Lars Vilhelm Henschen*, p. 105.
44. *Folkbladet*, August 19, 1849.
45. Gladh, *Lars Vilhelm Henschen*, p. 172.
46. Lundqvist reports their presentation at some length in *Erik Jansismen*, pp. 50 f. See Jonas Olsson's own account of the trial in *Erik-Janssonisternas historia*, pp. 36 ff.
47. Gladh, *Lars Vilhelm Henschen*, p. 171.
48. "Ett besök hos Erik Jansson på lazaretet," *Gefleborgs Läns Tidning*, December 7, 1844.
49. *Hudikswalls Veckoblad*, March 8, 1845. A correspondent wrote to *Norrlands-Posten* and commented on the irony of seeing a papermaker throwing books into the fire (April 1, 1845).
50. *Stora Kopparbergs Läns Tidning*, December 24, 1844.
51. *Hudikswalls Veckoblad*, March 8, May 10, and May 31, 1845.
52. Gladh, *Lars Vilhelm Henschen*, p. 182.
53. *En natt vid Bullarsjön* is no doubt the best known of the novels inspired by Janssonism. But there were others: C. A. Wetterbergh, *Pastorsadjunkten* (1845); J. L. Stockenstrand, *Utvandrare* (1907); Per Nilsson-Tannér, *Det nya Eden* (1934); Wiktor Norin, *Drömmen om Kanaan* (1950), and *Folket vid floden* (1952); Stuart David Engstrom, *They Sought for Paradise* (1939); Reuben H. Nelson, *Little Nels and the Partner* (1965). Sven O. Bergquist, *Fågelvägen* (1959), *Rästställen* (1960), and *Gränsmarker* (1961).

7. MOUNTING FURY

1. "Söderala Kyrkorådsprotokoll," Härnösand Landsarkiv.
2. The statement was signed by Jonas Olsson, Olof Olsson, Olof Nilsson of Berga, Nils Hedin (the tailor from Söderala), and Anders Jonsson of Lynäs.
3. Uppsala Domprotokoll, April 30, 1845 (Uppsala Landsarkiv). Two other Janssonists, Olof Jonsson of Oppsjö and Lars Persson of Delsbo, were also called before the Cathedral Chapter on August 27 and given official warnings.
4. Gladh, *Lars Vilhelm Henschen*, p. 183.
5. When some of this parishoners left for America the following year, Pastor Arenander made some angry comments in the Österunda *Utflytningsbok:* "Anders Pehrsson: book burner, father hater, tore apart the catechism." "Pehr Jansson: scorns Luther's teaching." He also noted that Pehr had been punished for making sounds during the church service, but we do not know if he snored or made snorts of disapproval (Uppsala Landsarkiv).
6. April 22, 1845. His release was also noted by *Gefleborgs Läns Tidning*, April 19, 1845, and by *Aftonbladet*, April 22, 1845.
7. Olof Stoneberg, as he called himself in America, became a trustee of the colony (Olsson, *Swedish Passenger Arrivals*, pp. 244–45). Later he joined the Methodist church. A letter written by him is in the library of Bethel Seminary, Stockholm.
8. He had recently moved from Forsa to Torstuna (*Hudikswalls Veckoblad*, April 12, 1845).

9. J. E. Ekblom, "Läseriet i Österunda och Eric Janssonismen, åren 1843–1846, samt anhängar i del i Sverige och Amerika," Anna Lindewall Collection, Uppsala Landsarkiv.

10. July 12, 1845.

11. The doors have been put on display in the lobby of the police headquarters in Hudikswall.

12. *Erik-Jansonisternas historia*, pp. 58–59.

13. *Hudikswalls Veckoblad*, July 12, 1845.

14. Helena Lindewall, "Om Erik Jansismen i Forssa, 1845," MS in the Herlenius Collection, Uppsala University Library.

15. Olof Stenberg's father, Jonas, was known in the neighborhood as "Skojar-Jon" (Swindler John). ("Husförhörslängd, 1846," Härnosand Landsarkiv).

16. J. E. Ekblom, "Läseriet i Österunda," in "Fröken Anna Lindewall Samling," Uppsala Landsarkiv.

17. Ibid.

18. Gladh, *Lars Vilhelm Henschen*, p. 218, n. 8.

19. *Erik-Jansonisternas historia*, p. 26.

20. "Inreligande handlingar," in Härnosand Landsarkiv.

21. Lundqvist, *Erik-Jansismen i Helsingland*, p. 186.

22. L. W. Henschen, *Kyrka och stat* (Uppsala, 1850), p. 6, n. 1.

23. *Hudikswalls Veckoblad*, November 8, 1845.

24. G. E. Klemming and J. G. Nordin, *Svensk boktryckeri-historia, 1843–1883* (Stockholm, 1883), p. 570.

25. *Gefleborgs Läns Tidning*, December 20, 1845; *Erik-Jansonisternas historia*, p. 61.

26. *Hudikswalls Veckoblad*, November 8, 1845.

27. The first trial was to be held at Sanna, and a large crowd had gathered there on October 11. But the trial was postponed until October 30 (*Hudikswalls Veckoblad*, Oct. 18, 1845).

28. *Erik-Jansonisternas historia*, pp. 70–74. J. O. Nordendahl reported to Baron Lagerheim that Pehr Pehrsson, his wife, his son, three daughters, and an unmarried couple who sometimes visited the Pehrssons were the only Janssonists in the area (Gladh, "Till hälsingeläseriets karakteristik," p. 204).

29. S. Sjöholm, "Två rapporter från kyrkoherden Lars Landgren till ärkebiskopen C. F. af Wingård om Erik-Jansismen," *Kyrkohistorisk årsskrift* 11 (1910):154.

30. *Hudikswalls Veckoblad*, November 1, 1845.

31. Ibid., March 4, 1848. The three men were Erik Olsson, Vitterarv; Olof Olofsson, Källeräng; and Pehr Larsson, Oppsjö.

32. Sjöholm, "Två rapporter," pp. 156–57.

33. *Hudikswalls Veckoblad*, November 8, 1845.

34. Ibid., November 22, 1845.

35. *Erik-Jansonisternas historia*, pp. 30–31. In a personal letter to me dated June 27, 1974, Ture Karlström, archivist for the city of Gävle, said that the castle prison until 1846 had cells which were like caves, without windows and with entrances through a hole in the roof. It was no doubt the prospect of spending the rest of his days in such an inhospitable setting that terrified Eric Jansson, who had been imprisoned in Gävle, and knew what he was in for.

36. *Hudikswalls Veckoblad*, December 3, 1845.

8. AMERICA HAS IT BETTER

1. *Erik-Jansonisternas historia*, pp. 41–43.

2. This account of Jonas Olsson's visit to Stockholm is based on his own recollections as he recorded them in "Eric Jansson's Following in Sweden," a chapter in *Erik-Jansonisternas historia*, pp. 35–53.

3. Arfwedson had written a dissertation on the New Sweden Colony while at Uppsala University. He was the author of *The United States and Canada in 1832, 1833, and 1834* (London, 1834), and served as the American consul in Stockholm during the years 1838–55 (Olsson, *Swedish Passenger Arrivals*, p. 9).

4. Ibid., p. 68. *Östgöta Correspondenten* reported their departure on October 4, 1845.

5. Most of the material in this paragraph comes from Emil Herlenius, "Erik-Jansismen i Dalarne," *Meddelanden från Dalarnes Fornminnenförening* 9 (1924):8–9.

6. *Tidning for Falu Län och Stad*, February 5, 1846.

7. *Stora Kopparbergs Läns Tidning*, August 7, 1847. Though the Janssonists seemed to have had a distinctive appearance in Sweden, they had no special uniform in America, and were visibly indistinguishable from their neighbors. The women of the colony wore blue drill cloth for workdays, and gingham or calico for Sundays and special occasions (Mikkelsen, *The Bishop Hill Colony*, p. 55).

8. Herlenius, "Erik-Jansismen i Dalarne," p. 9. The sisters were Maria Sofia Bouvin and Sara Helena Bouvin.

9. Gladh, "Till Hälsingeläseriets och Erik-Jansismens karakteristik," p. 197.

10. *Några sånger*, pp. 59–61. See also hymns number 9, 18, 35, 70.

11. Ekblom, "Läseriet i Österunda," p. 54. On at least two occasions, Jansson was examined by physicians who suspected paranoia or mental unbalance, but he was never certified as insane. He sometimes boasted of these challenges to his sanity, because he said they proved his claim to be Christ's suffering emissary (*Gefleborgs Läns Tidning*, December 7, 1844).

12. Ibid., April 15, 1846; and Upsala, April 17, 1846.

13. *Norrlands-Posten*, October 17, 1845.

14. MS written by Linjo Jonas Jonsson, a nephew of Linjo Lars Gabrielsson, in the Herlenius Collection, Uppsala University Library. See Emil Herlenius, *Ur Dalarnes kultur och personhistoria: Religiösa rörelser i Malung under mediet av 1800-talet* (Falun, 1934), pp. 5–7.

15. A damaged copy of this letter is in the archives of the Bishop Hill Heritage Association, Bishop Hill, Illinois.

16. *Afskedstal* (Galva, Ill. [1880]). The publisher claimed that the book was "printed from the original manuscript."

17. Ibid., pp. 2–3.

18. Ibid., p. 4.

19. Eric Johnson, "Swedish Colony at Bishop Hill, I," *Viking* 1 (March 1907):10. The painting by Krans which often has been identified as Eric Jansson's party leaving Oslo in a small boat is not that subject at all. It is Krans's copy of a familiar nineteenth century engraving by Clarence N. Dobell called "From Shore to Shore" (*Galvaland*, May 1970).

20. *Stora Kopparbergs Läns Tidning*, August 27, 1846.

21. Ibid., June 27, 1846.

22. *Gefleborgs Läns Tidning*, October 14, 1846.

23. A family story told to me by Fru Gunvor Jansson of Skensta, Fjärundra, Västmanland.

24. *Hudikswalls Veckoblad,* September 26, 1846.

25. In the *Österunda Husförhörslängd* [Household Examination Book], Arenander wrote after Jan Andersson's name, "Book burner" (Uppsala Landsarkiv).

26. Herlenius Collection, Uppsala University Library.

27. Enoch Berlin, "Forlåtelse utan gräns," *Julhälsning till Forsa församling* (1954), pp. 16–17.

28. *Upsala,* June 5, 1846.

29. *Gefleborgs Läns Tidning,* October 14, 1846.

30. "Landskonsliet i Gefleborgs Län, 1825–70," Härnosand Landsarkiv.

31. Ibid.

32. *Stora Kopparbergs Läns Tidning,* May 7, 1846.

33. Olsson, *Swedish Passenger Arrivals,* p. 68.

34. The original letter, which is in the archives of the Bishop Hill Heritage Association, was translated by Wesley Westerberg and Nils Runeby, and was published in the *Swedish Pioneer Historical Quarterly* 23 (April 1972):60–70.

35. *Hudikswalls Veckoblad,* November 8, 1845.

36. Witting, *Minnen från mitt lif som sjöman,* pp. 60–62.

37. *Tidning for Falu Län och Stad,* September 23, 1847.

38. Herlenius, *Ur Dalarnes kultur,* p. 7.

39. Olsson, *Swedish Passenger Arrivals,* pp. 86–98.

40. *Gefleborgs Läns Tidning,* August 8, 1846. Jansson wrote in his "Autobiography" that he met Margareta Hebbe in Söderala in 1843. She was later to be of great help to him at Bishop Hill (Olsson, *Swedish Passenger Arrivals,* p. 109).

41. *Stora Kopparbergs Läns Tidning,* June 18, 1846; *Hudikswalls Veckoblad,* November 14, 1846.

42. Theodore Schytte, *Vägledning för emigranter. En kort framställning . . . med ett bihäng om de år 1847 utvandrade Erik Janssons anhängares sorgliga öde* (Stockholm, 1849), pp. 40–46. John Norton identified "the Norwegian doctor."

43. August 2, 1846.

44. *Norrlands-Posten,* October 23, 1846; and *Hudikswalls Veckoblad,* October 23, 1846.

45. The leader on the *New York* was Anders Berglund, later a chief administrator for the colony, and later still a Methodist minister. Also aboard was Jonas Danielson, the colorful wandering minstrel who was known by the Bishop Hill people as "Doodle" (*Galvaland,* August, 1970).

46. Andrew Barlow (earlier Anders Berglöf) told Philip Stoneberg that the latter's ancestor, Olof Stoneberg (earlier Stenberg) had bought the bark *Aeolus,* which was loaded with iron (notes dated October 1, 1907, Knox College MS).

47. My information on the Janssonist ship movements comes largely from Olsson, *Swedish Passenger Arrivals,* passim, and from a list prepared by Carolyn A. Wilson, sometime student at Knox College.

48. *Tidning for Falu Län och Stad,* July 9, 1846.

49. J. E. Ekblom, *Beskrifning öfver Thorstuna Socken* (Enköping, 1872), p. 11.

50. See John Norton, "And Utopia Became Bishop Hill," pp. 7, 23. It should be pointed out that Arfwedson was genuinely interested in America and was not solely concerned with making money.

51. The excellence of the artifacts and architecture of Bishop Hill, has lent sub-

stance to the frequently made assumption that the colonists were not ordinary Swedish immigrants, but came from a higher class. However, careful studies made by Professor Charles Nelson, chairman of the Department of Sociology-Anthropology at Muskingum College, New Concord, Ohio, have shaken this view. His research demonstrates that one-third of the colonists were indeed peasants (landholders), but they were apparently the least prosperous of this class. Two-thirds of the colonists were cotters, crofters, and servants, the classes at the bottom of the Hälsingland totem pole ("Towards a More Accurate Approximation of the Class Components of the Erik Janssonists" [thesis, Muskingum College, 1974]).

52. F. Thorelius, *Försök till fullständiga upplysningar om läseriet* (Stockholm, 1855), p. 8.

53. The letters are in the Uppsala Landsarkiv under the heading "Handlingar ang. Eric Janssons irrlära, 1846." The clergy who thought that the *läsare* did in fact lay the foundation for the Janssonist movement are A. A. Scherdin (Söderala), T. E. Tjerneld (Övanåker), Otto Söderlund (Bollnäs), J. P. Haggberg (Gnarp), Eric Norlinder (Bergsjö and Hassela), J. G. Elfsberg (Skog), Olof Ericsson (Norfoss), Hans Egnell (Arbrå and Undersvik), L. O. Berg (Norrbo and Bjuråker). The clergy who defended the *läsare* movement, and said they were not to be blamed for the Janssonists, were P. N. Lundquist (Maråker), N. G. Wallden (Gävle and Valbo), C. L. Boström (Hudikswall), Eric Bengt Schilling (Forsa), Lars Landgren (Delsbo), Albert Steinmetz (Rogsta and Ilsbo), Olof Bolund (Hedesunda), V. Vadman (Norrala), A. Norell (Trönö), A. E. Ronquist (Mo and Regnsjö), Gabriel Lindstrom (Hamrånge), L. Rosengren (Voxna), J. O. Nordendahl (Enånger and Njutånger).

54. Cited by Söderblom, "Läsare och amerikafarare på 1840-talet," p. 86. See T. E. Tjerneld, "Några ord om läseri och Jansismen," *Norrlands-Posten*, February 24, 1846, and February 27, 1846.

55. Cited by Nils Runeby, *Den nya världen och den gamla: Amerikabild och emigrationsupfattning i Sverige, 1820–60* (Uppsala, 1969), pp. 214–15.

56. Ibid., p. 217.

57. *Ostgöta Correspondenten*, May 16, 1846.

58. *Norrlands-Posten*, July 12, 1849.

59. *Stora Kopparbergs Läns Tidning*, April 9, 1846.

60. Eric Johnson and C. F. Peterson, *Svenskarne i Illinois* (Chicago, 1880), p. 233. Carl Magnus Flack has been confused sometimes with his brother, Gustaf Flack (Olsson, *Swedish Passenger Arrivals*, p. xvi).

61. "Flack and Åstrom are in Chicago," (*Gefleborgs Lans Tidning*, October 14, 1846).

62. Letter of January 6, 1847, cited by Gladh, *Lars Vilhelm Henschen*, p. 117.

63. *Hudikswalls Veckoblad*, November 22, 1845.

64. See. John Norton, "Robert Baird, Presbyterian Missionary to Sweden of the 1840's," *Swedish Pioneer Historical Quarterly* 23 (July 1972):151–67; and Henry M. Baird, *The Life of the Rev. Robert Baird* (New York, 1866). See also Franklin D. Scott, "American Influences in Norway and Sweden," *Journal of Modern History*, 18 (March 1946):37–47.

65. Cited by Runeby, *Den nya världen och den gamla*, p. 215.

66. Ibid., p. 468.

67. The song is thought to have been inspired by the last shipload of Janssonists to leave for America, which sailed in 1854 (Ella Odstedt, "Erikjansismen i Nordångermanlåndsk folktradition," *Forum Theologicum* 19 [1962]:63–73).

9. LAND OF MILK AND HONEY

1. It is difficult to give a precise figure for the Janssonist colony. Professor Michael A. Mikkelsen estimated that there were between fifteen hundred and four thousand Janssonists in Sweden, of whom eleven hundred left for America (*The Bishop Hill Colony*, pp. 11, 21). Philip J. Stoneberg, a descendant of the colonists who was fascinated by their history, thought that eleven hundred emigrated (*100 Years. A History of Bishop Hill, Illinois*, ed. Theodore J. Anderson [Chicago, 1946], p. 29). Olov Isaacson and Soren Hallgren set the figure at twelve hundred (*Bishop Hill: A Utopia on the Prairie* [Borås, 1969], p. 53). Ulf Beijbom thought that there might be as many as fifteen hundred (*Swedes in Chicago: A Demographic and Social Study of the 1846–1880 Immigrants* [Växjö, 1971], p. 48). The difficulty is that poor records were kept, and that the numbers varied. Some Janssonists never left Sweden, some left the sect on the way at New York or Chicago, or after they had arrived at Bishop Hill, some died, and some were born on the way. Twelve hundred seems like a round, uncertain number which is as satisfactory as any.

2. Olsson, *Swedish Passenger Arrivals*, pp. 68–69. The son, Jonas, had infantile paralysis, and had to be left at home in Sweden. He made his own crossing in the fall of 1846, traveling in the company of his grandfather and his aunt, and arrived in Bishop Hill the day after his mother died.

3. Westerberg and Runeby, "Document: A Letter from Olof Olsson," pp. 61–70. A good account of Olof Hedström is in Ewald Benjamin Lawson, "Olof Gustaf Hedström," *American Swedish Historical Museum Yearbook* (Philadelphia, 1945), pp. 63–74.

4. *Galva Weekly News*, October 1, 1896, p. 3.

5. See N. M. Liljegren et al., *Svenska Metodismen i Amerika* (Chicago, 1895), p. 166.

6. *Helsi*, April 3, 1847. Blombergsson was assigned the task of meeting the Janssonist ships which arrived in New York during the summer of 1846.

7. Olson, *The Swedish Element in Illinois*, pp. 50–51.

8. Eric Johnson, "Fifty Years in Politics," *Viking* 1 (July 1906):11.

9. John E. Lilljeholm, *Pioneering Adventures of John Edvard Lilljeholm in America, 1846–1850*, trans. Arthur Wald, Augustana Historical Society, Publications, vol. 19 (Rock Island, Ill., 1962), p. 14. Many years later, Mrs. Hellström remembered that "he was to have married Mrs. Pollock" (Conversation with Philip Stoneberg, Knox College Collection).

10. *Erik-Jansonisternas historia*, pp. 51–53. Anders Larsson wrote about the affair to Bailiff Ekblom. He said that Ljungberg, who had been a sailor, was hired by the Janssonists because he could speak English. Ljungberg's story was that the Janssonists took not only their own money, but his as well, and for good measure took his clothes from the chest. Anders Andersson, Jonas Berglund, and Jonas Ericsson were temporarily jailed, but they were released when Ericsson proved that Ljungberg really owed him four thousand dollars (Fröken Anna Lindewall Samling, Uppsala Landsarkiv).

11. Otto Stenberg, *Erik-Jansismen i Nord-Amerika* (Söderhamn, 1848), p. 5. Stenberg wrote that "if I knew when I was in Sweden what I know now, I would never have come to America" (p. 3).

12. Westerberg and Runeby, "Document: A Letter from Olof Olsson," p. 67. See also Schytte, *Vägledning för emigranter*, pp. 41–44.

13. Johnson and Peterson, *Svenskarne i Illinois*, p. 27. One of the authors, Captain Johnson, is the son of Eric Jansson.

14. Anders Larsson, juryman from Hällby, Österunda Parish, wrote seventeen letters to the bailiff J. E. Ekblom from Chicago between 1846 and 1865. They are now in the Uppsala Landsarkiv. Some of them were published in Albin Widén, *När Svensk-Amerika grundades. Emigrantbrev* (Borås, 1961). The household examination book from Österunda contains the note that Anders Larsson was once tried on a charge of forgery but was acquitted.

15. Beijbom, *Swedes in Chicago*, p. 45.

16. Ibid., pp. 45–46.

17. *Najaden*, November 14, 1846.

18. Witting, *Minnen från mitt lif som sjöman*, pp. 86–87.

19. Gustaf Unonius, *Pioneer in Northwest America* (Uppsala, 1862), 2:203.

20. Johnson, "Swedish Colony at Bishop Hill," p. 18.

21. Letter to P. Wieselgren, May 23, 1850, published in Gunnar Westin, *Emigranterna och kyrkan* (Stockholm, 1932), p. 43. The Lutheran pastor, G. Palmquist, had the same view. Writing to Norelius from Galesburg on January 9, 1852, he said, "Wesley himself, the founder of Methodism, does not teach what Hedström does here" (Erik Norelius, *De svenska luterska församlingarnas och svenskarnes historia i amerika* [Rock Island, Ill.], 1890, p. 154).

22. Johnson, "The Swedish Colony at Bishop Hill," pp. 18–19. Witting said that Olof Olsson gave his forty-acre farm to Jansson and the colony (*Minnen från mitt lif som sjöman*, pp. 176–77, n. 8).

23. Letters to Sweden spelled the local place names with wild abandon, sometimes accidentally, sometimes in fun. We read "Oppal" and "Upphal" for Hopole, and the two groves were sometimes called "Sodom and Gomorrah." See the letter from Eric Erickson, written in November, 1847, and published in Bror Hillgren, *En bok om Delsbo* (Stockholm, 1925), p. 76.

24. George Swank, *Bishop Hill, Showcase of Swedish History: A Pictorial History and Guide* (Galva, Ill., 1965), p. 8.

25. He lived on the first stairway, north side, on the third floor, if Helena Lindewall remembered rightly. (See Mrs. O. Pilstrand to Philip J. Stoneberg, October 2, 1907, Knox College MS.)

26. Ibid., p. 9. A German, August Bandholz, married into the community and became the chief mason during the building of Big Brick.

27. Farmers in Henry County at this time grew corn, wheat, oats, and rye—in that order (U.S. Bureau of Census, *Compendium of the Seventh Census, 1850*, pp. 228–29; and *Eighth Census*, Agriculture, pp. 30–37).

28. *Nytt brev från Amerika om Eric Jansarnes tillstånd derstädes* (Söderhamn, 1850), p. 4.

29. Arthur Eugene Bestor, Jr. *Backwoods Utopias: The Sectarian and Owenite Phases of Communication Socialism in America, 1663–1829* (Philadelphia, 1950), p. 22.

30. By 1856 the colonists seem to have softened this policy. When Baron Axel

Adelswärd visited them in that year he found them "very good to the poor Swedes who came to them. If they wish they may join them, but if not they may stay with them for some time and are fed and housed free" (H. Arnold Barton, ed., *Letters from the Promised Land: Swedes in America, 1840–1914*. [Minneapolis, 1975], p. 81).

31. Charles Nordhoff, *The Communistic Societies of the United States: From Personal Visits and Observations* (1875; reprint ed., New York, 1960), pp. 83 ff. Because of the bitterness of those leaving, the principle was changed in 1836, and after that recovery of investment, sometimes was offered "at the discretion of the superintendent" (ibid., p. 84).

32. Communalism seems a better term than communism, which brings up irrelevant images of Marxist revolutionaries.

33. Cited Widén, *När Svensk-Amerika grundades*, p. 43.

34. *Nytt bref från Amerika*, p. 5. Lonborg surely was mistaken when he said that Jansson recommended that men should satisfy their sex urges in public: "He said that if a man had desire for his wife, they should fall down where they were immediately and have intercourse. He said, 'A person should never be ashamed to carry out God's will before men.' "

35. Paulus Magnus Myrtengren had been married to Johanna Maria Kleen, who had died (Olsson, *Swedish Passenger Arrivals*, p. 97).

36. Carrie Lindbeck to Philip J. Stoneberg, the Stoneberg Collection, Knox College.

37. Stoneberg, *100 Years*, p. 41.

38. *Aftonbladet*, April 17, 1848. Jansson himself stressed to his followers that they should never be dependent on any one man; but he did not like outsiders stressing the point.

39. Not exactly true. The 1850 census records the presence at Bishop Hill of at least one Tennessee family: Haydn and Lucinda Hilton and their three children, Tom, John, and Elmira.

40. Nordhoff, *Communistic Societies*, p. 212. See also *History of Henry County*, p. 137, and Herman Richard Mueller, *Fighter for Freedom* (New York, 1959), pp. 63–79.

41. Anders Blomberg had been held for a year at Central Hospital, Uppsala, where he was treated for being mentally disturbed (*Hudikswalls Veckoblad*, September 20, 1845, and *Upsala*, April 17, 1846).

42. Esbjörn met him there in 1851 (Olsson, *Swedish Passenger Arrivals*, p. 54).

43. Nordhoff said that a seventy-year-old Shaker woman said to him, "I say to any who know me, as Jesus said to the Pharisee, 'Which of you convicteth me of sin?' " (*Communistic Societies*, p. 63). The Oneida perfectionists thought that the Second Coming had happened in 70 A.D., and since that time individuals could be entirely free from sin—provided that they lived according to "Bible communism" (Marion Lockwood, "The Experimental Utopia in America," *Daedalus* [Spring, 1965], p. 403). The official Shaker name was "The United Society of Believers in Christ's Second Coming."

44. Philip J. Stoneberg Collection, Knox College.

45. Interview with Eric Benson on March 26, 1903. There were two periods of celibacy at Bishop Hill, the first ending in June, 1848, when Jansson decided that marriage was part of God's plan, and the second period beginning in 1854, when Nils Hedin came back from a visit to the Shaker communities and convinced Jonas

Olsson that celibacy was not only more pleasing to God, but also led to affluence (Raymond Lee Muncy, *Sex and Marriage in Utopian Communities: Nineteenth-Century America* [Bloomington, Ind., 1973], p. 32). Little thought seems to have been taken of the psychological havoc such decisions could make. Jonas Westlund protested bitterly that because of this teaching, his own wife had attempted suicide (Herlenius, *Erik-Jansismens historia*, pp. 133–34). And Anna Maria Stråle reported to a friend that shortly after 1854, Jonas Olsson's own wife had committed suicide by drowning in a well (cited by Widén, *När Svensk-Amerika grundades*, p. 49). Anders Blomberg returned to Sweden in 1867 as a Shaker, calling himself "Doctor." In Älvdalen he persuaded sixty-three people to emigrate to the Shaker colony at Pleasant Hill (Emil Herlenius, *Religiösa rörelser i Falun under mediet av 1800-talet* [Falun, 1942], pp. 11–12). Blomberg was head of the "family" of Swedes in the West Lot Family House, but the Swedes lost their enthusiasm for the Shakers, and the West Lot House was closed in 1882. (Marywebb Gibson, *Shakerism in Kentucky* [Cynthiana, Ky., 1934] p. 79).

46. Mikkelsen, *Bishop Hill Colony*, p. 31.

47. Cited by Vilhelm Moberg, *Den okända släkten* (Stockholm, 1950), p. 29.

48. *Hudikswalls Veckoblad*, July 17, 1847.

49. Cited by Newman, *Gemenskaps och frihetssträvanden*, p. 277, n. 48.

50. Cited by Swank, *Bishop Hill*, p. 30.

51. Johnson and Peterson, *Svenskarne i Illinois*, pp. 98–99.

52. Johnson, "Fifty Years in Politics," *Viking* 1 (July 1906):12. Anders Larsson mentioned politics to Ekblom in a letter dated March 7, 1848: "Germans, Norwegians, Swedes and Danes are for the most part Democrats, and their intention this time is to keep Irishmen out of office" (letter in Widén, *När Svensk-Amerika grundades*, p. 90).

10. PORTENTS OF DISASTER

1. "One of the [Bishop Hill] pilgrims told me that young and strong as she was, had there been a chance to set foot in any land, whatever its horrors or inhabitants, nothing could have persuaded her back into the ship" (Emma Farman, "A Plymouth of Bishop Hill and Its Founder Eric Janson," *The American-Scandinavian Review* 2 [September, 1914]:32).

2. Olson, *Swedes in Illinois*, p. 269.

3. Lilljeholm, *Pioneering Adventures of Johan Edvard Lilljeholm*, pp. 20–34.

4. Unonius, *A Pioneer in Northwest America*, 2:208.

5. Norelius, *De Svenska Luterska församlingarnas historia*, p. 136. But Norberg must not have been on unfriendly terms with the other colonists. He came back to help defend the colony when the Root mobs attacked in 1850. In 1851 he was in Minnesota, and is thought to have been the discoverer of Chisago Lake, the setting for Moberg's epic trilogy. Back in Bishop Hill in 1852, he played a prominent role as secretary during the legal trials of "the Colony Case." In 1867 he and his wife, Brita Jansdotter, were living with their two children in Toulon, Illinois, where he died in 1899.

6. But he was present, at least, when the Janssonists burned books at Tranberg on June 19, 1844 (*Gefleborgs Läns Tidning*, June 19, 1844). And Janssonist conventicles were held in his home at Bollnäs (B. J. Jonzon, *Blad ur Bollnäs församlings kyrkohistoria* [Uppsala, 1899], p. 182).

7. *Norrlands-Posten,* March 1, 1849. The eccentric Bäck may have lived like a hermit, but if so, it was like a rich one. He claimed to have the biggest house between Chicago and Galesburg (Johnson and Peterson, *Svenskarne i Illinois,* p. 155).

8. L. P. Esbjörn wrote to Dr. Badger on February 28, 1850, that the Janssonist deserters were not easy to live with. "We have among us, especially in Galesburg, many former Janssonists who have left Bishop Hill, but still are Janssonists in their hearts, still perfectionists (in the spirit of Joseph Smith), still full of pride and selfishness, still love any preacher who preaches perfectionism, and still suspicious of any minister who prays for forgiveness of sins" (Norelius, *De Svenska Luterska församlingarnas historia,* p. 133).

9. J. S. Chambers, *The Conquest of Cholera, America's Greatest Scourge* [New York, 1938], pp. 85–118.

10. Patrick E. McLean, "The St. Louis Cholera Epidemic of 1849," *Missouri Historical Review* 63 (January 1969):171–81. *The Chicago Daily Democrat* on January 16, 1850, gave a harrowing report of the St. Louis epidemic.

11. Ramon Powers and Gene Younger, "Cholera on the Overland Trails, 1832–1869," *Kansas Quarterly* 5 (Spring 1973):32–49.

12. Norelius, *De Svenska Luterska församlingarnas historia,* p. 117.

13. Lilljeholm, *Pioneering Adventures of Johan Edvard Lilljeholm,* p. 41.

14. Norelius, *De Svenska Luterska församlingarnas* historia, p. 29. Cleng Peerson, "the pathfinder of Norwegian immigration," at the age of sixty-five joined the colony, contributed all his money, married a Swedish girl, and shortly afterward left (Theodore C. Blegen, "Cleng Peerson and Norwegian Immigration," *Mississippi Valley Historical Review* [March 1921], pp. 323, 328).

15. Johnson and Peterson, *Svenskarne i Illinois,* p. 88. A monument marking the site was erected in 1888, near the Cambridge road (Stoneberg, *100 Years,* p. 42).

16. Johnson and Peterson, *Svenskarne i Illinois,* p. 106.

17. Powers and Younger, "Cholera on the Overland Trails," pp. 32–49.

18. Witting, *Minnen från mitt lif som sjöman,* p. 194.

19. Unonius, *A Pioneer in Northwest America,* 2:206. See also N. M. Liljegren et al., *Svenska Metodismen i Amerika,* p. 169.

20. Unonius, *A Pioneer in Northwest America,* 2:206.

21. Londborg, *Nytt bref från Amerika,* p. 6. The letter was also printed in *Helsi* (Söderhamn) on September 9, 1849. See also Unonius, *A Pioneer in Northwest America,* 2:208.

22. The son, Jonas, was taken into the home of his uncle, Peter Dalgren, who lived in Galesburg (Johnson and Peterson, *Svenskarne i Illinois,* p. 299).

23. The above details are from Olsson, *Swedish Passenger Arrivals,* and from Londborg, *Nytt bref från Amerika,* pp. 5–6.

24. Witting, *Minnen från mitt lif som sjöman,* p. 98.

25. *Aftonbladet,* April 17, 1848.

26. Londborg, *Nytt bref från Amerika,* p. 5.

27. Ibid., p. 4.

28. August 11, 1849, and August 25, 1849.

29. He was accidentally killed by an earth slide on November 26, 1856. His wife then married a colonist named Silén (Olsson, *Swedish Passenger Arrivals,* p. 89).

30. Westin, *Emigranterna och kyrkan,* p. 44. It is noteworthy that Rönnegård, who searched the Esbjörn sermons for the period 1843–50, could find no example

of a direct attack against Jansson. The closest, perhaps, is Esbjörn's remark at Septuagesima, 1845, that the Christian life "is not perfect, without any fault—Paul testifies about himself—his own and other apostles' examples" (Rönnegård, *Lars Paul Esbjörn*, p. 66). Esbjörn did say of Jonas Hedström: "This man, who is in a terrible battle with Erik Jansson, preached about the same perfectionism as Jansson" (*Norrlands-Posten*, June 20 and 27, 1850).

31. Philip J. Stoneberg notes of conversation with Mrs. H. Lindewall, October 17, 1907 (Knox College MS).

32. Herlenius, *Erik-Jansismens historia*, pp. 76 f.

11. THE SHOOTING

1. From a letter from Britton A. Hill to Governor Augustus C. French, published in Henry E. Pratt, "The Murder of Eric Janson, Leader of the Bishop Hill Colony," *Journal of the Illinois State Historical Society* 45 (1952):64. Attorney Hill became something of an expert on fiscal policy, and wrote three books on the subject: *Liberty and Law under Federal Government* (Philadelphia, 1873); *Absolute Money: A New System of National Finance* (St. Louis, 1875); and *Specie Resumption and National Bankruptcy* (St. Louis, 1876).

2. Olsson, *Swedish Passenger Arrivals*, p. 97. Fine portraits of Caroline and her husband, Pehr Erik Ersson, are in the Swedish Pioneer Historical Society archives.

3. Johnson and Peterson, *Svenskarne i Illinois*, p. 184. The Hedstrom MS says that six men arrived together: "John Root came to Bishop Hill in 1847, accompanied by four Swedish men, Er[ic] Wäster, Håkanson, Captain Simmerman from Falun, Löfgren, and Norberg. These men tried very hard to marry girls from Bishop Hill, but only Routh, Håkanson, and Norberg succeeded" (fol. 1). The information is questionable. Captain Zimmerman was born in Västerås in 1800, but he was a captain in the Dalarna regiment. If "Norberg" is Eric U. Norberg, the latter came to Bishop Hill from Michigan. They may of course have arrived together.

4. The French connection for these three adventurers must be labeled a frontier yarn, but Zimmerman, at least, served both in the Swedish and the United States Army. Charles Zimmerman, as he called himself in this country, joined the Swedish army at the age of fifteen, and reached the rank of captain by 1828. He resigned from the Swedish army in 1840 and made his way to America. Enlisting in the United States Army in New York in 1847, he fought in Company B of the Seventh Infantry Regiment during the Mexican War, and was discharged with the rank of sergeant at Jefferson Barracks, Missouri, on July 31, 1848. (K. A. Leijonhufvud, *Ny Svensk släktbok*, *The Army Register of Enlistments for the Mexican War*, and the Adjutant General's Office, National Archives, Washington, D.C. I owe this information to Dr. Nils W. Olsson.)

5. Unonius, *A Pioneer in Northwest America*, 2:320.

6. Ibid., p. 321. It is said that he moved to Wisconsin, and when he came back to Princeton in 1850 he was too poor to pay the freight on his barber stool. He opened a general store in Princeton, selling cigars, whiskey, and clothing, and did so well that he soon had to open a branch store in Galesburg. In 1856, he moved to Dallas, Texas, and so drops out of our picture (Johnson and Peterson, *Svenskarne i Illinois*, pp. 184–85).

7. E. Gustav Johnson, "A Prophet Died in Illinois a Century Ago," *Swedish Pioneer Historical Quarterly* 1 (1950):8–12.

8. A search of the register for enlistments in the Mexican War, kept in the National Archives, Washington, D.C., has produced no evidence of John Root having served in that war. If he fought in the war at all, it is likely that he used another name.

9. "Settler," who seems to have had access to primary sources, said that he had to leave Sweden because of "an act akin to forgery" (*Henry County Chronicle*, February 28, 1860). See also Johnson and Peterson, *Svenskarne i Illinois*, p. 39.

10. Lotta Root gives this date in her petition for divorce, filed in the Henry County Courthouse, Cambridge, Illinois, in May 1853.

11. The original document is in the archives of the Bishop Hill Heritage Association. Lotta was illiterate and usually had John Hällsén to attest her mark.

12. Professor Mikkelsen may have heard this story from Jonas Olsson, with whom he had long conversations (*Bishop Hill Colony*, pp. 38–39). The Hedstrom MS repeats the story, but says the man who disappeared was a Norwegian (fols. 2).

13. Pratt, "The Murder of Eric Janson," p. 57.

14. Johnson and Peterson, *Svenskarne i Illinois*, p. 40.

15. Unonius, *A Pioneer in Northwest America*, 2:209. Unonius spoke of Root as "one of the most dangerous, most savage, and crafty persons I have ever known" (ibid., p. 211).

16. Herlenius, *Erik-Jansismens historia*, p. 75.

17. This story lends some credence to Root's claim that he served in the Mexican War, since he would then have been eligible for a veteran's bounty. On July 9, 1847, Anders Larson wrote to Ekblom: "I've spoken with a Swede from Skåne, who has just returned from Mexico, and who has taken part in remarkable battles. He has been paid $120, and been given 160 acres of land for his year in the Army." (John Norton, ed., "For it Flows with Milk and Honey," *Swedish Pioneer Historical Quarterly* 24 [July 1973]:178.)

18. John was born in January 1849, according to the petition of divorce filed in the Henry County Courthouse by Lotta Root. The Census of 1850 must be mistaken when it gives the birth date as March 12.

19. Pratt, "The Murder of Eric Janson," pp. 57–58.

20. May 25, 1850.

21. Norelius, *Early Life*, p. 111.

22. Unonius lived near Per Ersson's house, and said that he had talked with the Root family at that time: "As far as I was able to judge, the wife was planning to leave her husband and return to Bishop Hill. She secretly informed Jansson of her whereabouts and he sent a messenger to Chicago with whom, according to my belief, she went back to the colony willingly and without her husband's consent" (Unonius, *A Pioneer in Northwest America*, 2:209–10).

23. *Norrlands-Posten*, July 26, 1850; and *Gem of the Prairie*, May 25, 1850.

24. *Mormonism and American Culture*, ed. Marvin S. Hill and James B. Allen (New York, 1872), pp. 76 f.

25. Hedstrom MS, fol. 8.

26. Pratt, "The Murder of Eric Janson," pp. 55–69.

27. After Jonas's death, the diary went to Mrs. Maud Bercer of Los Angeles. It is now in the Illinois State Historical Society, Springfield. See Karin Ankarberg,

"Några avsnitt ur Bishop-Hill koloniens historia," *Historiska studier tillägnade Folke Lindberg, 27 August 1963*, ed. Gunnar Westin et al. (Stockholm, 1963), pp. 123–31.

28. Ernest W. Olson, "Relikvarium: Fornt och färskt ur vår Amerikahistoria," *Ungdomsvännen* (1917), pp. 16–19.

29. Johnson and Peterson, *Svenskarne i Illinois*, p. 87.

30. Olson, "Relikvarium," pp. 16–17.

31. Ibid. The answers given to questions 53, 55, and 56 in Jansson's *Catechism* assert that it is proper for a Christian to lie if the survival of the true faith depends on such untruth (pp. 58–59). The quotation is an important clue as to Jansson's attitude toward Root, but my translation is not certain. The words Jansson used, according to Olson's transcription, are these: "Jag har sagt att ingen har ti na mig så mycket som Rut." *Ti na* is gibberish, so I have changed the word to *pinat*.

32. MS letter in the Illinois State Historical Society Archives, Springfield, Illinois.

33. Ibid., p. 18.

34. Mikkelsen, *Bishop Hill Colony*, p. 42; and Swank, *Bishop Hill*, p. 39.

35. The wooden building in which Jansson was shot was used as a courthouse for thirty years. In 1878, it was moved to the S. W. Krapp farm, five miles west of town, and served as a barn until it was torn down in 1952 (*Corn, Commerce, and County Living*, ed. Terry Ellen Palsen [Moline, Ill., 1968], p. 66).

36. For the details of the shooting I have leaned heavily on the Hedstrom MS, fols. 12–13. Jansson's last words were, "En sugg ware god nog till hustru åt Root!"

12. WAKING FROM THE DREAM

1. *Galena Daily Advertiser*, May 23, 1850. The Henderson MS quotes Root saying, "I've done a crazy thing. You can do whatever you want with me" (fols. 12–13).

2. The True Bill is in the archives of the Henry County Courthouse, Cambridge, Illinois.

3. Timothy Hopkins, *The Kelloggs in the Old World and the New* (San Francisco, 1903), 2:1222.

4. Kenneth A. Williams, "William Kellogg of the Illinois Fourth District: Lincoln's Friend" (Master's diss., Illinois State University, 1968), p. 17.

5. *Gem of the Prairie*, May 25, 1850.

6. It may have been true that Jansson at some time or other threatened Root. It is certainly true that he used fear as an instrument of control over the colony and cultivated the notion that he had the power of life and death. But it is surely not true, as Root must have known, that Jansson meant to follow up such threats with physical violence.

7. MS in the Henry County Courthouse, Cambridge, Illinois.

8. *Henry County Chronicle*, March 20, 1860.

9. The present marble monument marking the grave has an inscription in Swedish which time has effaced, but which fortunately was copied by a visitor, J. Sällström, a Mission Friend minister from Galesburg, in 1904: "En minnesvärd af en älskad fader och make af Erik, Mathilda (son och dotter), Anna Sophia (enka) samt erkänsla af undertecknade till en vän, den vi högaktade medan han lefde och

vars gods och lärdomd vi aldrig förglömma." ("A memorial to a beloved father and husband from Erik and Mathilda [son and daughter] and Anna Sophia [widow] and also a pledge by the undersigned to a friend whom we highly honored while he was alive, and whose good and true teachings we shall never forget.") It was signed by Johan and Wilhelm Westberg, Hans and Marta Nordström, Erik and Marta Nordström, Anna Maria Stål, Anna Hedberg, Olof Olsson, and Erik Hedberg" (see *Missions-Wännen*, August 30, 1904). Probably there are some errors in the pastor's translation rather than on the monument: "Wilhelm" should be "Wilhelmina," and "Stål" should be "Stråle." There is also an inscription in English on the other side of the monument: "Erik Janson/Founder/of the town of/Bishop Hill/Born in Biskopskulla/Sweden/Dec. 19, 1808,/Murdered/May 13, 1850."

10. Letter dated May 21, 1850. Cited by Sam Ronnegård, *Utvandrarnas kyrka* (Stockholm, 1961), p. 88.

11. May 25, 1850.

12. MS letter in the Illinois State Historical Library, Springfield, Illinois.

13. Fredrica Bremer happened to be coming down the Mississippi at this time, and on November 2, stopped over night at Rock Island. Some Janssonists came aboard and invited her to attend the trial of Root, which was to begin the next day, but she begged off. The nights were cold, she said, she did not feel well, and besides, why should she go? (*Hemmen i den nya verlden* [Stockholm, 1853], 2:364.)

14. They were James McCully, James Brown, James McCord, James McPherrin, Simeon Gibbs, John Perdue, Henry Cooper, Asaph de Long, Anthony Beard, Robert C. Brown, Green Butterfield, and Judson W. Randall.

15. "Settler," in *Henry County Chronicle*, March 13, 1860.

16. Document in the Henry County Circuit Court, Cambridge, Illinois.

17. Ibid.

18. Unpublished letter, Illinois State Archives, Springfield, Illinois. The letter was called to my attention by James Oliver of Divernon, Illinois. Walsh's letter was endorsed by Marcus N. Osborn. Benjamin Walsh moved to Rock Island in 1850, after the cholera epidemic, and opened a successful lumberyard. Later he achieved some distinction as an entomologist. He died in 1869, after a Chicago and Rock Island railway train had run over his foot. See Edna Armstrong Rucker, "Benjamin D. Walsh—First State Entomologist of Illinois," *Transactions of the Illinois State Historical Society for the Year 1920* (Springfield, 1921), 27:54–61. Walsh edited the first volume of the *Transactions of the Illinois State Horticultural Society for the Year 1867* (Chicago, 1868).

19. The letter is in the Illinois State Archives, Springfield, Illinois, and was also found by James Oliver. It was signed by Lydia Matteson, Julia M. Rossiter, Delsie A. Payne, Mrs. Goodall, and the governor's wife, Mary Fish Matteson.

20. Unonius, *A Pioneer in Northwest America*, 2:211. "Settler" was under the impression that Root practiced as a lawyer in Chicago, but there is no confirming evidence.

21. March 16, 1856.

22. Cited by Witting, *Minnen från mitt lif som sjöman*, p. 242. Erik Shogren (or Sjögren as it was spelled in Sweden) was born in Gnarp, Gävleborg Province, in 1824, and came to America together with a party of Janssonists on the *Solide* in 1849. He assisted Pastor Hedström for a while on the Bethel ship in New York harbor, and then served various Methodist churches in Illinois and in California,

where he died in 1906 (Olson, *Swedish Passenger Arrivals*, p. 207). He was the pastor of the Bishop Hill Methodist Church from 1870 to 1873.

23. The affidavit was written for her by her sister, Anna Marie Janson, and to it she affixed her mark, with Johan Hällsén serving as witness. The manuscript is in Bishop Hill Heritage Association archives. Johan Hällsén was in his later years a writer of Adventist tracts. He spoke at the sixtieth anniversary of the Bishop Hill Colony in 1906, and died in 1913 at the age of ninety-three.

24. Stoneberg, *100 Years*, p. 49; Thomas D. Clark and F. Ferald Ham, *Pleasant Hill and Its Shakers* (Pleasant Hill, Ky., 1968), p. 62. In 1873 Will Carleton wrote the well known "Over the Hill to the Poor House" about the poorhouse where Sophia died (George Swank, *Historic Henry County* [Galva, Ill., 1941], p. 25).

25. Johnson and Peterson, *Svenskarne i Illinois*, p. 298.

26. Barton, *Letters from the Promised Land*, p. 81.

27. Johnson and Peterson, *Svenskarne i Illinois*, p. 312.

28. Bishop Hill was not officially a colony until this charter was granted.

13. A LIVING MUSEUM

1. Cited by Arthur Sundstedt, *Striden om konventiketplakatet* (Stockholm, 1958), pp. 239–40. Archbishop Wingård called Anders Sandberg an *acer vindex* (sharp defender) of the Janssonists (Sjöholm, "Två rapporter," p. 155).

2. Sundstedt, *Striden om konventiketplakatet*, p. 242. Thomander was a strong defender of freedom for the orthodox. "Surely it is barbaric and bloodthirsty teaching," he said to Parliament, "that religious sanctity should be sustained by the executioner's axe!" (cited by Nils Algård, *Johan Henrik Thomander, kyrkomannen—personlighet* [Stockholm, 1924], p. 328).

3. Newman, *Gemenskaps och frihetssträvanden*, pp. 280 ff., and Lars Osterlin, *Thomanders kyrkogärning* (Lund, 1960), p. 406.

4. Sundstedt, *Striden om konventiketplakatet*, p. 242, and Gladh, *Lars Vilhelm Henschen*, p. 304.

5. Moberg, *Den okända släkten*, p. 29. Moberg said that he admired Jansson's skill as an administrator, but it is more likely Jansson's attack against the church administration which won Moberg's approval.

6. Rönnegård, *Utvandrarnas kyrka*, pp. 91–92.

7. See Franklin D. Scott, "Sweden's Constructive Opposition to Emigration," *Journal of Modern History* 37 (September 1965):307–35. A useful bibliography of the Swedish reaction to the emigration is in Esther Larson, *Swedish Commentators on America, 1638–1885* (New York and Chicago, 1963).

8. *The Social Teaching of the Christian Churches*, trans. Olive Wyon (New York, 1931), 1:331.

9. Max Weber, *The Theory of Social and Economic Organization*, trans. A. M. Henderson and Talcott Parsons (New York, 1947), p. 358.

10. *The Basic Writings of C. G. Jung*, ed. Violet Laszlo (New York, 1959), p. 541.

11. "November 2, på Mississippi," *Hemmen i den nya verlden* (Stockholm, 1853), 2:245.

12. P. Waldenström, *Genom norra Amerikas förenta stater* (Chicago, 1890), p. 386. Professor George Stephenson made the same point more harshly: "Ignorance

and overweening self-confidence combined with stupid persecution, drove Jansson to absurd extremes" (*The Religious Aspects of Swedish Immigration* [Minneapolis, 1932], p. 50).

13. Anna Söderblom, *En Amerikabok*, p. 197.

Selected Bibliography

Åkerlund, P. A. *Om kyrkans och statens enhet.* Stockholm, 1854.

Alstermark, Bror. *De religiöst-svärmiska rörelserna i Norrland 1750–1800. I. Herjeådalen och Helsingland.* Strängnäs, 1898.

Anderson, Theodore. *100 Years: A History of Bishop Hill.* Chicago, 1946.

Andersson, Albert. "Erik-Jansismen—ett 100-års minne." *Hudiksvall Tidning,* May 31, 1944.

Andersson, S. "Något om Erik Jansismen, särskilt dess verksamhet i Alfta." *Julhälsning till församlingarna i ärkestiftet, 1923,* pp. 86–100. Uppsala, 1923.

Ankarberg, Karin. "Några avsnitt ur Bishop-Hill koloniens historia." *Historiska studier tillägnade Folke Lindberg, 27 augusti, 1963,* edited by Gunnar T. Westin, pp. 123–33. Stockholm, 1963.

Baird, Henry. *The Life of the Rev. Robert Baird.* New York, 1866.

Baird, Robert. *A Visit to Northern Europe.* 2 vols. New York, 1841.

———. *Religion in America.* New York, 1844.

Beijbom, Ulf. *Swedes in Chicago: A Demographic and Social Study of the 1846–1880 Immigration.* Translated by Donald Brown. Växjö, 1971.

Benson, Adolph B. and Naboth Hedin. "The Bishop Hill Colony." *Americans from Sweden,* pp. 106–18. Philadelphia, 1950.

Berg, C. W. and Amy Moberg. *Teckning af Carl Olof Rosenii lif och werksamhet, hans wänner tillegnad.* Stockholm, 1868.

Berggren, John Erik. *Om den kristliga fullkomligheten.* Uppsala, 1887.

Bestor, Arthur Eugene, Jr. *Backwoods Utopias: The Sectarian and Owenite Phases of Communitarian Socialism in America, 1663–1829.* Philadelphia, 1950.

Bigelow, Hiram. "The Bishop Hill Colony." *Transactions of the Illinois State Historical Society for the year 1902,* pp. 101–8. Springfield, Ill., 1902.

Bishop Hill Colony Case, The. Answer to the bill of complaint, filed in the Henry circuit court, to the October term, A.D. 1868. Motion to dissolve the injunction, and brief and parts, on motion of Bennett & Veeder, solicitors, and of counsel for the defendant. Galva, Ill., 1868.

Blegen, Theodore C. "Cleng Peerson and Norwegian Immigration." *Mississippi Valley Historical Review* 7 (March 1921): 303–31.

Bohlin, Torsten. *Lars Landgren, människan, folkuppfostraren, kyrkomannen.* Stockholm, 1942.

Bonham, Jeriah. *Fifty Years' Recollection.* Peoria, Ill., 1883.

Bremer, Fredrika. "November 2, på Mississippi." *Hemmen i den nya verlden.* Stockholm, 1853.

Brodin, Knut, ed. *Emigrantvisor och andra visor.* Stockholm, 1938.

Calkins, Earnest Elmo. *They Broke the Prairie: Being Some Account of the Upper Mississippi Valley by Religious and Educational Pioneers.* New York, 1937.

Calverton, Victor Francis. *Where Angels Dared to Tread*, pp. 117–26. Indianapolis and New York, 1941.

Chaiser, John. *Genmäle af John Chaiser öfver de tre Artiklarna i Sändebudet som författats utaf J. A. Gabrielson, A. J. Anderson, och B. A. Carlson.* Bishop Hill, 1874.

Cohn, Norman. *The Pursuit of the Millennium.* London, 1970.

Cornelius, C. A. *Svenska kyrkans historia efter Reformationen.* Uppsala, 1887.

Dearinger, Lowell A. "Bishop Hill Colony. Rise and Fall of Swedish Dissenters Communal Society on the Illinois Prairie." *Outdoor Illinois* 4 (July 1965): 4–13.

E——m, Karl. "Erik Jansismen." *Swenska Weckobladets Månadsupplaga*, no. 1 (January 1887).

Ekman, E. J. *Den inre missionens historia*, pp. 802–24. Stockholm, 1898.

En sång angående Erik Jansista Willfarelsen, författad af en Bonde i Norra Helsingland. Hudiksvall, 1846.

Erdahl, Sivert. "Eric Janson and the Bishop Hill Colony." *Journal of the Illinois State Historical Society* 17 (October 1925): 503–74.

Erik-Jansonisternas historia. Galva, Ill., ca. 1902.

Erikson, J. M. *Metodism i Sverige. En historisk-biografist framställning af Metodistkyrkans verksamhet i vårt land.* Stockholm, 1895.

Fant, G. T. and A. T. Låstbom, eds. *Upsala ärkestifts herdaminne.* Vols. 1–3. Uppsala, 1842–45.

Farman, Emma Shogren. "A Plymouth of Swedish America: The Town of Bishop Hill and Its Founder, Eric Janson." *The American Scandinavian Review* 2 (September 1914): 30–36.

Fredenholm, Axel. "100-årsminnet av Janssonismen." *Hemåt*, pp. 100–105. Chicago, 1946.

Gladh, Henrik. "Till hälsingeläseriets och Erik-Jansismens karakteristik." *Kyrkohistorisk årsskrift* 47 (1947): 186–212.

———. *Lars Vilhelm Henshen och religions-frihetsfrågan till 1853.* Uppsala and Stockholm, 1953.

H., U. *Nysätra och Österunda.* 1843.

Hallsén, John. *Religionens frihet från lag.* Stockholm, 1895.

Hamlin, Talbot. *Greek Revival Architecture in America*, pp. 307–9. Oxford, 1944.

Hatten, Minnie Maxwell. "Bishop Hill: A Sectarian Community." Master's thesis, State University of Iowa, n.d.

Havighurst, Walter. "From Helsingland to Bishop Hill. *Upper Mississippi: A Wilderness Saga.* New York, 1944.

Hedblom, Folke. "Hos hälsingar i Bishop Hill: från en Amerikaresa, 1962." *Hälsingerunor*, pp. 12–20. Sundsvall, 1963.

Hedin, Naboth, and Adolph B. Benson. "Svensk bosättning i Illinois: kolonien Bishop Hill." *Vår Svensk stam på utländsmark: I Västerled.* Gothenburg, 1952.

Henschen, L. W. *Kyrka och stat.* No. 2. Uppsala, 1850.
Herlenius, Emil. *Erik-Jansismen i Sverige.* Uppsala, 1897.
———. *Erik-Jansimens historia: Ett bidrag till kännedom om det Svenska sektväsendet.* Jonköping, 1900.
———. "Erik-Jansismen i Dalarne." *Meddelanden från Dalarnes Fornminnesförening* 9 (1924): 7–22.
Hinds, William Alfred. *American Communities and Co-operative Colonies.* Chicago, 1908.
Holloway, Mark. "Racial and Religious Communism." *Heavens on Earth: Utopian Communities in America, 1680–1880.* New York, 1951.
Isaksson, Olov. "Discover Bishop Hill." *Swedish Pioneer Historical Quaterly* 19 (October 1968): 220–33.
Isaksson, Olov, and Sören Hallgren. *Bishop Hill: A Utopia on the Prairie,* translated by Albert Read. Stockholm, 1969.
Jacobson, Margaret E. *Bishop Hill 1846.* Bishop Hill, Ill., 1941.
———. "The Painted Record of a Community Experiment: Olof Krans and His Pictures of the Bishop Hill Colony." *Journal of the Illinois State Historical Society* 34 (June 1941): 164–76.
Janson, Florence K. *The Background of Swedish Immigration, 1840–1930.* Chicago, 1931.
Jansson, Eric. *En härlig beskrifning på menniskans tillväxt, då hon helt är, enligt Johannes 15 cap., inympade uti det sanna winträdet Christo, eller en sann förklaring öfver 2 Konungs boken 2: huru den som lika med Elisa hafver öfvergifwit allt, kan tillwara under troende bön, till att tala dubbelt så mycket, som sin mästare. Ett ord i sinom tid till läsaren.* Printed by C. G. Blombergsson, Söderhamn, 1846.
———. *Ett afskedstal, till alla Sveriges innevånare, som har föraktadt mig, den Jesus hafver sändt; eller förkastat det namnet Erik Jansson, såsom orent, för det jag har bekänt 'Jesu namn' inför menniskar.* Printed by C. G. Blombergsson, Söderhamn, 1846. 2d ed. Galva, Ill., 1902.
———. *Ett ord i sinom tid, eller en kort wederläggning af "Erik Jansismen i Helsingland."* Printed by C. G. Blombergsson, Söderhamn, 1846.
———. *Förklaring öfver den Helige Skrift, eller catechės författad i frågor och svar.* Printed by C. G. Blombergsson, Söderhamn, 1846.
———. *Några ord till Guds församling.* Printed by C. G. Blombergsson, Söderhamn, 1846.
———. *Några sånger samt böner.* Printed by C. G. Blombergsson, Söderhamn, 1846. 2d ed. (with additional hymns) Galva, Ill., 1857.
Johansson, Olof. *Bland främder och främlingar.* Stockholm, 1926.
Johnson, E. Gustav. "A Prophet Died in Illinois a Century Ago." *The Swedish Pioneer Historical Quarterly* 1 (July 1950): 8–12.
Johnson, Eric. "Sixtieth Anniversary of the Settlement of Bishop Hill, Henry County, Illinois." *Viking* 1 (November–December 1906): 4–5, 6–9.
———. "Swedish Colony at Bishop Hill, I." *Viking* 1 (March 1907): 9–10, 18–19.
Johnson, Eric and C. F. Peterson. *Svenskarne i Illinois.* Chicago, 1880.
Jonsson, Falkens Per. *Tillståndet hos de till Amerika utvandrade Erik-Jansarne samt profetens bedragerier, skildrade i bref från en utvandrare.* Gävle, 1847.
Jonsson, Ingvar. "Fäbodbebyggelsen i Hälsingland." *Hälsingerunor, 1964–1965,* pp. 157–66. Malung, 1964.

Jonzon, B. G. *Blad ur Bollnäs församlings kyrkohistoria.* Uppsala, 1899.

Killgren, Bror. *En bok om Delsbo.* Stockholm, 1925.

Kiner, Henry L. *History of Henry County, Illinois.* 1: 621–51. Chicago, 1910.

Klefbeck, Alarik. *Etiska ideer i svensk frikyrklig väckelsereligiösitet.* Stockholm, 1952.

Klemming, G. E. and J. G. Nordin. *Svensk boktryckerihistoria, 1483–1833.* Stockholm, 1883.

Krantz, Magni. "Om Erik Jansson och hans anhängare." *Hudiksvalls Tidning,* 1950: March 11; July 13; July 24; August 1, 18, 21; September 4, 11, 25; October 3, 31; November 13, 28. (The Krantz articles were reprinted in *Svenska Amerikanaren Tribunen:* March, October, November, 1950.)

Lagerberg, Matt. "The Bishop Hill Colonists in the Gold Rush." *Journal of the Illinois State Historical Society,* no. 48 (Winter 1955), pp. 466–69.

Lagergren, Anders. *Om de andliga rörelserne i Sverige, särdeles med afseende på den riktning separatisterna utvecklat i Norrland och Dalarne.* Stockholm, 1855.

Laing, Samuel. *A Tour in Sweden in 1838 Comprising Observations on the Moral, Political, and Economic State.* London, 1839.

Lamm, M. *Upplysningtidens romantik: Den mystiskt sentimentala strömmingen i Svensk litteratur, 1–2.* Stockholm, 1918–20.

Landgren, L. "Om de antinomistiska rörelserna i Helsingland." *Tidskrift för Svenska kyrkan,* pp. 319–37. Uppsala, 1849.

Liljegren, N. M.; N. O. Westergreen; and C. G. Wallenius. "Bishop Hill." *Svenska Metodismen i Amerika,* pp. 200–203. Chicago, 1895.

Lilljeholm, John E. *Pioneering Adventures of John Edward Lilljeholm in America 1846–1850.* Translated by Arthur Wald. Augustana Historical Society Publications, vol. 19 (1962).

Linder, Oliver A. "The Story of Illinois," *The Swedish Element in America.* 1: 37–47. Chicago, 1931.

Lindstrom, David E. "The Bishop Hill Settlement." *Yearbook 1945 of the American Swedish Historical Museum,* pp. 55–62. Philadelphia, 1945.

Londberg, Daniel. *Nytt bref från Amerika om Erik Jansarnes tillstånd derstädes.* Söderhamn, 1850.

Lowe, D. G. "Prairie Dream Recaptured." *American Heritage* 20 (October 1969): 14–23.

Lundqvist, P. N. *Erik-Jansismen i Helsingland. Historisk och dogmatisk framställning jemte wederläggning af läran.* Gävle, 1845.

Mikkelsen, Michael A. *The Bishop Hill Colony: A Religious Communistic Settlement in Henry County, Illinois.* Johns Hopkins University Studies in Historical and Political Science, ser. 10, no. 1. 1892. Reprint. Philadelphia, 1972.

Minnen från Voxna-bygd. Bollnäs, 1928.

Moberg, Vilhelm. *Den okända släkten.* Stockholm, 1950.

Morton, Stratford Lee. "Bishop Hill: An experiment in Communal Living." *Antiques* 43 (February 1943): 74–77.

Nelson, Helge. "Bishop Hill—the First Swedish Settlement in Illinois." *The Swedes and the Swedish Settlements in North America.* Lund, 1943.

Nelson, Ronald E. "The Bishop Hill Colony and Its Pioneer Economy." *Swedish Pioneer Historical Quarterly* 18:32–48.

———. "Bishop Hill; a Colony of Swedish Pietists in Illinois." *Antiques* 99 (January 1971): 140–47.

Newman, E. *Gemenskaps och frihetssträvanden i Svensk fromhetsliv, 1809–1855*. Lund, 1939.

Newton, J. A. "Perfection and spirituality in the Methodist tradition." *Church Quarterly* 3 (October 1970): 95–103.

Nohrborg, Anders. *Den fallna människans salighetsordning, föreställd uti betraktelser öfwer de årliga sön-och högtids-dagars evangelium*. Stockholm, 1771.

Nordhoff, Charles. *The Communistic Societies of the United States: From Personal Visit and Observation*, pp. 343–49. 1875.

Nordquist, Del. "Olof Krans: Folk Painter from Bishop Hill, Illinois." *American Swedish Historical Foundation Yearbook, 1961*, pp. 45–58. Philadelphia, 1961.

Norelius, Erik. *De Svenska Luterska församlingarnas och Svenskarnes historia i Amerika*. Rock Island, Ill., 1890.

———. "The Swedish Background of the Settlement of Bishop Hill, Illinois." *The Covenant Quarterly* 11 (August 1951): 83–90.

Norton, John E. "Bishop Hill Rediscovered." *Augustana Swedish Institute Yearbook, 1968–69*.

———. "And Utopia Became Bishop Hill." Thesis, Augustana College, December 15, 1971.

———. "Robert Baird, Presbyterian Missionary to Sweden of the 1840's." *Swedish Pioneer Historical Quarterly* 23 (July 1972): 151–67.

———, ed. " 'For It Flows with Milk and Honey,' Two Immigrant Letters about Bishop Hill." *Swedish Pioneer Historical Quarterly*, 24 (July 1973): 163–79.

Ohlsson, Johan. "Om Erik Jansismen och kolonien Bishop Hill." *Hälsingerunor* (1964–65), pp. 81–95.

Olson, Ernest W. "The Bishop Hill Colony." *History of the Swedes of Illinois*." 1: 197–270. Chicago, 1908.

———. "The Bishop Hill Colony." In Gösta Nyblom, *Americans of Swedish Descent: How They Live and Work*, pp. 124–31. Rock Island, Ill., 1948.

———, ed. "Erik Jansons epistlar till de Bishop-Hillare." *Ungdomsvännen* 22 (January 1917): 16–19.

Olsson, Nils William. *Swedish Passenger Arrivals in New York, 1820–1850*. Chicago, 1967.

———, ed. *A Pioneer in Northwest America, 1841–1858: The Memoirs of Gustaf Unonius*. Translated by Jonas Oscar Backlund. 2 vols. Minneapolis, 1950, 1960.

Osterlin, Lars. *Thomanders kyrkogärning*. Lund, 1960.

Peterson, C. F. *Svenskarna i Illinois*. Chicago, 1880.

———. "Bishop Hill: Var Svensk-Amerikanska moderkoloni." *Valkyrian*, 2 (June 1898): 300–304.

———. *Sverige i Amerika*. Chicago, 1898.

Pratt, Dorothy and Richard. "Bishop Hill (1848–1860)," *A Guide to Early American Homes, North*. New York, 1956.

Pratt, Harry Edward. "The Murder of Eric Janson, Leader of the Bishop Hill Colony." *Journal of the Illinois State Historical Society* 45 (1952): 55–69.

Qvarnström, Axel. *Anteckningar om Söderala socken i Gefleborgs län*. Söderhamn, 1904.

Reuterdahl, Henrik. *Ärkebiskop Henrik Reuterdahls memoarer*. Lund, 1920.

Rimbe, Set. "Frälsaren på Stenbo." *Julhälsning till Forsa församling* (1954) pp. 8–13.

Rönnegård, Sam. *Lars Paul Esbjörn och Augustana-synodens uppkomst.* Stockholm, 1949.

———. *Utvandrarnas kyrka: En bok om Augustana.* Stockholm, 1961.

Runneby, Nils. *Den nya världen och den gamla.* Uppsala, 1969.

S., O. *Erik-Jansismen i Nordamerika.* Söderhamn, 1848.

Sandewall, Allan. "Konventikelplakatets upphavande—ett gransår i svensk religionsfrihetstafstiftning?" *Kyrkohistorisk årsskrift* 58 (1957): 137–52.

Schytte, Theodore. *Vägledning för emigranter.* Stockholm, 1849.

Scott, George. *Sveriges religiösa tillstånd sådant det blivit uppfattadt och uti Amerika framstäldt.* Stockholm, 1841.

Sefstrom, A. G. *Några blad till historien om läsarne med afseende fästadt på de inom Helsingland vistande.* Falun, 1841.

Semi-centennial Celebration of the Settlement of Bishop Hill Colony, Held at Bishop Hill, Illinois. Galva, 1909.

Serenius, A. "Svenska minnen i nya världen." *Hemåt,* pp. 61–74. Chicago, 1918.

"Seventieth Anniversary: Celebration of the Seventieth Anniversary of the Founding of the Bishop Hill Colony." *Journal of the Illinois State Historical Society* 9 (October 1916): 344–56.

Sewerin, John. "Erik-Jansarna i Hälsingland: Också ett 100 årsminne av Johan Sewerin." *Ljusdalsposten,* July 19, 1942.

Sjöberg, Sven. "Profeten från Bishop Hill och hans öde," *Upplands Forminnesförenings årsbok.* Uppland, 1950.

Sjöholm, F. "Två rapporter från kyrkoherden Lars Landgren till ärkebiskop C. F. af Wingard om Erik Jansismen." *Kyrkohistorisk årsskrift* 11 (1910): 154–60.

Sjöstrand, Martin. "Det svåra året." *Hyltén-Cavalliusföreningens årsbok* (1937), pp. 7–24.

Skarstedt, Ernest. "Bishop Hill kolonien i Illinois," *Svenska-Amerikanska folket i helg och socken.* Stockholm, 1917.

Skogsbergh, E. August. "Bishop Hill väckelsen." *Minnen och upplevelser.* Minneapolis, 1925.

Söderblom, Anna. "Ett besök i Bishop Hill." *En Amerikabok.* Stockholm, 1925.

———. "Läsare och Amerikafarare på 1840-talet; brev, protokoll m.m. om Erik Jansismen." *Julhelg för Svenska hem,* pp. 80–93. Stockholm, 1925.

Spooner, Harry L. "Bishop Hill: An Early Cradle of Liberty." *The American Scandinavian Review* 45 (March 1959): 29–38.

[Stenberg, Otto.] *Erik-Jansismen i Nord-Amerika, eller beskrifning om Erik-Jansarnes tillstånd derstädes, samt resan dit med skeppet New York. Bref från en af utvandrarne.* Söderhamn, 1848.

Stephenson, George M. "Astrology and Theology." *Swedish-American Historical Bulletin* 2 (August 1929): 53–69.

———. "Eric-Jansonism and the Bishop Hill colony." *The Religious Aspects of Swedish Immigration.* Minneapolis, 1932.

Stockenstrand, J. L. "Erik Jansarne." *Utvandrare.* Stockholm, 1907.

Stoneberg, Philip J. "Bishop Hill koloniens industriella lif," *Vinter-rosor 1908.* Chicago, 1908.

———. "The Bishop Hill Colony." *The History of Henry County, Illinois* by Henry L. Kiner. Chicago, 1910.

———. "The Bishop Hill Colony." *The Swedish Element in Illinois: Survey of the Past Seven Decades.* Edited by Ernest W. Olson. Chicago, 1917.

———. "The Bishop Hill Colony and the Notes of the Western Exchange Fire and Marine Insurance Co." *Numismatic* 30 (November 1917): 462–64.

Stråle, Anna Maria and Jonas Olson et al. *Erik-Jansonisternas historia.* Bishop Hill, 1884–94.

Sundstedt, Arthur. *Striden om konventikelplakatet.* Stockholm, 1958.

Swainson, John. "The Colony of Bishop Hill." *Scandinavia*, No. 2 (January 1883), p. 1.

———. "Swedish Colony at Bishop Hill, Illinois." *History of the Scandinavians and Successful Scandinavians in the United States*, edited by C. N. Nelson, pp. 135–52. Minneapolis, 1893.

Swank, George. *Bishop Hill: A Pictorial History and Guide.* Galva, Ill., 1965.

———. *Painter Krans O.K.* [*sic*] *of Bishop Hill.* Galva, Ill., 1976.

Tanner, Per Nilsson. *Det nya Eden: Ett Svenskt emigrations-äventyr.* Stockholm, 1934.

Unonius, Gustaf. "Besök hos Erik Jansonska kolonien; om koloniens uppkomst och utveckling." *Minnen från en sjutton-årig vistelse i Nordvestra Amerika.* 2: 384–86. Uppsala, 1862.

Upsala ärkestifts herdaminne. Edited by John E. Fant, August T. Låstbom, and Ludvig Nyström. Vols. 1–4. Uppsala, 1842–93.

Waldenström, P. "I Bishop Hill." *Genom Norra Amerikas Förenta Stater*, pp. 382–88. Chicago, 1890.

Webber, Everett. "Bishop Hill, Swedish Commune in Illinois." *Escape to Utopia: The Communal Movement in America*, pp. 274–77. New York, 1959.

Westerberg, Wesley M. "Bethel Ship to Bishop Hill." *Swedish Pioneer Historical Quarterly* 23 (April 1972): 55–59.

Westin, Gunnar. *Handlingar till George Scotts verksamhet i Sverige.* Uppsala, 1928, 1929.

———. *George Scott och hans verksamhet i Sverige.* 2 vols. Stockholm, 1929.

———. *Emigranterna och Kyrkan: Brev från och till svenskar i Amerika, 1849–1892.* Stockholm, 1932.

———. "Brev från L. P. Esbjörn, 1840–1850." *Kyrkohistorisk årsskrift* (1946), pp. 211–66.

Widen, Albin. "Bishop Hill: A Coming Centennial." *The American-Scandinavian Review* (September 1942), pp. 217–27.

———. *När Svensk-Amerika grundades.* Broås, 1961.

———. *Amerikaemigrationen i dokument.* Stockholm, 1966.

Wieselgren, H. "Jansson, Erik." *Svenskt biografiskt lexicon.* Örebro, 1863–64.

Wilson, Carolyn Anderson. "Revitalization, Emigration, and Social Organization: An Ethnohistorical Study of the Bishop Hill Colony, 1846–1861." Honors thesis, Knox College, 1973.

Wilson, Charles. *Några upplysande underättelser rörande Erik Jansonska kolonien i Norra Amerika: Samt en sarskild resebeskrifning.* Sundsvall, 1854.

Witting, Victor. *Minnen från mitt lif som sjöman, immigrant och predikant; samt en historisk afhandling af metodismens uppkomst.* Worcester, Mass., 1904.

Youngert, S. G. "Kolonien Bishop Hill." *Svenskarna i Amerika*, edited by Karl Hildebrand and Axel Fredenholm, 1:252–55. Stockholm, 1924.

Index